The Dolphin and Whale Career Guide

A complete sourcebook for anyone wanting to get involved with dolphins and whales.

by

Thomas B. Glen III

The Dolphin and Whale Career Guide:
A complete sourcebook for anyone wanting to get involved with dolphins and whales.

FIRST EDITION

Library of Congress Cataloging-in-Publication Data:
Glen, Thomas B. III
The Dolphin and Whale Career Guide: A complete sourcebook for anyone wanting to get involved with dolphins and whales. / by Thomas B. Glen III
p. cm.
Includes bibliographical references.

SF80.M55 1997
637.7'9400—dc20

90-50733
CIP

ISBN 0-9660675-6-8

1. Animal Specialists—Vocational Guidance 2. Careers 3. Dolphins, Training 4. Marine Mammals I. Title II. Title: The Dolphin and Whale Career Guide

CONTENTS

PART FOUR: APPENDICES

Acknowledgments

You always hear authors going on and on about all the help they've received from this person and that, and how their book could never have come to be without these people's help. I had no idea how true that is until I went through it myself.

My greatest thanks go out to Rachel Dixon, who has seen me through this project from beginning to end. I hardly ever see her, but she was always there for me, and bailed my butt out of Vancouver the first time I ran out of funds. Huge thanks also to Robin Heine, my editor and sometime conspirator. Robin donated far more time than she should have to help me pull all the bits together in time for Christmas. How we avoided being arrested while running amok in D.C. is beyond me.

I owe much to my aunt, Patricia Salberg. Aunt Pat took me in when I was doing the starving author thing, and put up with me for six months. I was able to gather the information in chapter four thanks to Ken Ramirez, who vouched for me wherever I went, and who assured the Navy I wasn't a raving madman. Cynthia Perkin did a tremendous job on the cover, and gave me much-needed advice on what looks good and what doesn't.

I'd like to thank Tim, Susan, Kris, Don, Tina, Vince, Kelly, and my Mom and Dad for believing in me. Your support meant more to me than any of you know.

A tip of the hat to Fred and Paul, who helped me survive Seattle. I was forced to work for a month in the engine room of the NOAA vessel MacArthur when I went broke the second time, and these two made an otherwise unpleasant situation a real blast.

Thank you Tamara and Mary, wherever you are. I left my heart in San Francisco. Twice.

Finally, thanks to all those who participated in this project, especially to those who donated photographs:

Cedar Point
Center For Coastal Studies
Coastal Ecosystems Research Fndn.
Dolphin Connection
Dolphin Research Center
Indianapolis Zoo
Kris Landin
Minnesota Zoo
Mirage
N.C. Maritime Museum

Part One

Introduction

- The Fascination of Dolphins and Whales
- A Wee Bit of Natural History
- How Smart Are They?

1
The Fascination of Dolphins and Whales

Why is everyone so jazzed up over whales and dolphins? They figure prominently in the myths and legends of dozens, maybe hundreds of cultures, including the Inuit tribes, the ancient Greeks and Romans, the Japanese and Chinese, and even within the earliest writings of Christianity. Each year thousands of people travel great distances to see these animals in the wild, and almost everyone on the planet would either pay to see them in an aquarium or fight to get them out. Dolphin and whale-related merchandise is a multi-million dollar industry in the United States, and the unfathomable success of movies like "Free Willy" attests to the enormous popularity of dolphins and whales in North America.

This is, of course, good news for me.

But what is it about these animals that captures our attention, our fascination, and our compassion? In the book West Coast Whale Watching, Richard Kreitman attributes it in part to the animal's sheer size, coupled with the knowledge that these leviathans are living, right now, in the Earth's oceans. Kreitman recalls standing awestruck under a life-size replica of a blue whale at New York's American Museum of Natural History. I know exactly which model he was referring to, since the very same one helped launch my own interest in dolphins and whales back in 1975. On that same visit, my father bought me one of those old stamp book things - they came with a couple of pages of picture stamps (they were way bigger than postage stamps) which you had to tear out, lick, and stick on the appropriate page. Once you got the picture in place, the text on the page told you about what you were looking at. The book my dad got me was all about whales, and I had it for years until it finally got lost with all the other beloved crap I really don't need taking up space anymore.

I doubt if the text of that booklet would be very accurate now, but it certainly sparked my imagination. Few things freaked me out like the drawing of the sperm whale in a titanic battle with a giant squid in the dark, murky depths of the ocean. For years, anything that I'd bump into while swimming was a giant squid. But my interest in dolphins and whales grew faster than my fears of the unknown. I loved going to the Shedd Aquarium in Chicago, and went as often as my parents would take me. I had posters of all the different species, read all I could find at the library, and even talked mom and dad into sending me to Seacamp on Big Pine Key, Florida, to get SCUBA certified at age 13. I watched Jacques Cousteau religiously, bought that Greenpeace record of humpback whale songs, and went swimming as often as I could. "The Man From Atlantis" was one of my favorite TV shows, and my dad would always welcome the two hours of peace and quiet he'd get whenever "The Incredible Mr. Limpet" was shown on Saturday afternoon.

I got hooked on the sea, just like thousands of kids are each year in this country. In high school, I resolved to go to a college with a good marine biology program so I could get a job working with marine mammals. I made a list of all the marine biology programs in the country, and wrote for application materials. Imagine my surprise when not one single college catalog even mentioned dolphins, whales, or any kind of marine mammal! At some point I started hitting up the local libraries for all their career information on the subject, and you know what I got? Butkis. Nothing. Nada, zip, zilch, zero, nil. I went to my guidance counselor: "Uhh..." I went to my Biology teacher: "Err..." Nobody had anything helpful to say at all, and I became pretty frustrated.

I strayed from the path after high school and joined the Army for a few years, but when I returned to Chicago, it wasn't long before my interest in whales brought me back to the Shedd Aquarium. The Aquarium built a new Oceanarium beside the old building in 1991, and I thought they might be able to use some help cleaning the tanks or something. I volunteered there for all of three weeks before I realized I'd found my next career. I spent as much time there as I could over the next nine months, often putting in twenty hours per week doing whatever odd jobs they'd let me. They finally hired me on as an animal care specialist, working with the Aquarium's marine mammals and penguins, and interacting with the visitors. Far and away the most common question I answered there was "how can I get a job like yours?" Each time someone asked me that, I was reminded of how fruitless my search had been in high school. I would spare as much time as I could to pass on what I had learned, but I never really felt I had answered the question adequately.

What you now hold in your hands is a slightly more complete answer. This book began in 1995 as a solution for students who wanted to become dolphin trainers. It wasn't long before I realized there were a lot of other ways to get involved with dolphins and whales, and I've done my best to cover them all.

The book is divided into four parts. Part One gives you a little background information to bear in mind while reading through the rest of the book. Part Two is the meaty part, and tells you everything you need to know about getting a job in dolphin and whale display, research, or conservation. Part Three is a resource guide both for career-seekers looking to increase their experience, or for anyone simply looking to spend their spare time around or studying dolphins and whales. The last section, Part Four, contains many useful appendices and organizational lists, including a listing of schools with marine mammal classes or programs.

There will be numerous additions to this text in the coming years. Indeed, the second edition was being laid out even before this book went to the printer. If you can, check out the Omega website periodically for updates to the Guide. (www.omegaland.com)

The most important thing to remember as you read on is that persistence, initiative, and dedication are mandatory for any of the fields covered herein. I can only point you in the right direction, and the rest is up to you. Now turn the page, and see what life has to offer...

2
A Wee Bit of Natural History

Okay kids, hang on tight, 'cause this goes pretty quickly. The purpose of this short section is to give you just enough background information to get started on the rest of the book. If most of this information is new to you, take a peek at the reading list on page 353. The books there can teach you almost anything you ever wanted to know about the biology, behavior, and history of dolphins and whales.

A few terms you need to know:

Baleen (bay LEEN) Baleen plates hang down from the upper jaws of baleen whales. The plates are fibrous, and are used to strain food from the water.

Cetacean (seh TAY shun) This term includes dolphins, whales, and porpoises, and is the word people use to refer to all three kinds of animals at the same time. "Cetacea" is the scientific order to which these animals belong.

Mysticete (MIST eh seet) The large whales with baleen instead of teeth. Mysticetes are often called "baleen whales". "Mysticeti" is one of two scientific sub-orders of Cetacea, the other being Odontoceti, below.

Odontocete (oh DAHNT eh seet) The "toothed whales", which includes beaked whales and all dolphins and porpoises. "Odontoceti" is the other scientific sub-order of Cetacea.

Pinnipeds (PIN uh peds) This describes animals which belong to the scientific sub-order "pinnipedia" which contains fur seals, sea lions, seals, and walrus. The term is often used to refer to such animals as a group.

Pseudorca (sood OARK uh) The genus name for a false killer whale, which is often used to refer to such an animal.

Tursiops (TERSE ee ops) The genus name for a bottlenose dolphin, which is often used to refer to such an animal.

Orca (OARK uh) The species name for a killer whale, which is often used to refer to such an animal.

What are cetaceans?

Cetaceans have been around for at least 30 million years. It's hard to tell just what kind of animals cetaceans evolved from, but the best evidence points to a family of smallish, fish-eating hoofed mammals which lived in North America, Europe, and Asia about 50 million years ago. Three main branches of whales developed, becoming the odontocetes and mysticetes we know today, and a now extinct line known as archaeocetes. There are now just under 80 recognized species of cetaceans which occur in almost every part of the world's oceans and in a number of rivers as well. Cetaceans are mammals, making them warm-blooded animals that give birth to and nurse their young, and breathe air like humans. The larger whales are some of the biggest animals on the planet, and the blue whale is, as far as anyone can tell, the largest animal which has ever lived on Earth.

The terms "dolphin", "whale", and "porpoise" are not scientific classifications, and occasionally get confused. All mysticetes are called whales, as are the larger odontocetes, like the sperm whale, beluga whale, and killer whale. Smaller odontocetes are usually called dolphins or porpoises, and the two terms are often used interchangeably. There are no universally accepted rules about when to use one term over another, but scientists tend to restrict the use of the term "porpoise" to the Phocoenidae family. A good rule of thumb is to call the smaller odontocetes with long, pronounced beaks “dolphins”, and those with no visible beak “porpoises”.

Cetaceans, by the way, are just one kind of marine mammal. Others include sea otters, polar bears, manatees and dugongs, and pinnipeds. All can be found in North American waters except dugongs, which hang out near Australia, Indonesia, and other warm parts of the South Pacific.

Dammit Jim, this is a career guide, not a biology textbook. If you still get confused about what dolphins and whales are, then hie thee to a library. Otherwise, read on...

3
How Smart Are They?

If you're over 25 years of age, chances are you've seen "The Day of The Dolphin". The 1973 film stars George C. Scott as a benevolent research scientist who teaches dolphins to speak English, only to have the military try to corrupt his work for its own sinister porpoises. (Har) The film was extremely popular and grounded in half-truths, which is always a dangerous combination in our society.

Dr. John Lilly, upon whom George Scott's character was loosely based, had indeed been attempting to "establish contact" with dolphins by deciphering "delphineese", a supposed dolphin language which would enable humans to chat with their aquatic neighbors as easily as foreign diplomats discussing policy. Dr. Lilly's research and credibility have since been ruined within the scientific community following a number of questionable experiments and conclusions. To date, he is the only researcher to have administered LSD to dolphins. Nevertheless, rumors of his work and the release of the film sparked the imagination of thousands. An entire generation of naive, enthusiastic potential scientists were instantly swept up in the notion that we could learn to speak to the dolphins.

This hypothesis was formed without due consideration of the facts involved and sent a number of capable researchers trudging blindly through lengthy and sometimes unproductive investigations in search of the "Northwest Passage" of the cetacean world. In the twenty-five years that followed, no one anywhere has "spoken" with a dolphin the way we speak with one another.

The past two decades have seen remarkable advances in cetacean research. There are droves of mysteries yet waiting for discovery, but a great deal is now known about the biology, social structure, and natural history of dolphins and whales. Much scientific data exists on the topic of cetacean intelligence. The issue is not now, and may never be settled, but it is possible to draw a number of conclusions from what data exists.

Brains, Brains, Brains

One of the most popular and enduring theories in favor of sentient dolphins cites their large brain size. Lots of people see the relatively large brains of dolphins and whales as evidence of higher intelligence. The truth is, absolute brain size has long been known to have little to do with higher intelligence. There are numerous animals of modest intellect with much larger brains than humans, such as rhinoceri, whale sharks, and elephants. Even within the human race there seems to be no relationship between brain size and intellect. ("Take men" says my editor. Hmpf.) In fact, Neanderthal man's cranium is larger on average than our own. Relative brain size also fails to support the concept of high cetacean intellect. In comparing the ratio of brain to body sizes, mice come out on top, scoring 50% higher than humans. Bottlenose dolphins run a close third, but most whales score beneath elephants, cows, and spiny anteaters.

What many people overlook is the difference between odontocete (toothed whales) and mysticete (baleen whales) body-brain ratios. Odontocetes have proportionately much larger brains than their baleen-bearing cousins. Apart from their dentition, the most remarkable difference between these groups is the odontocetes' use of echolocation, or biological sonar. This skill is incredibly advanced - far ahead of anything human science has been able to reproduce. Still largely mysterious, dolphin and whale echolocation is somewhat similar to the sonar used by bats, but it is much more accurate and versatile. Tests show that echolocation works as well as, and probably better than, visual sight for locating and identifying objects underwater. Even blindfolded dolphins can tell how far an object is, what direction it's in, what its dimensions are, what the texture of its surface is like, and even how dense the material is. Dolphins can correctly identify a three-inch sphere as far away as 650 feet, and have even detected manufacturing defects in supposedly identical hollow pipes. Is it any wonder animals with such an accurate and extensive sensory system would require a large brain to handle the extra bandwidth?

Another theory to explain large brains in cetaceans concerns their apparent lack of sleep. Researchers long ago discovered, to their horror, that administering general anesthesia to a dolphin is fatal, as breathing is an entirely conscious process for cetaceans. (Effective though seldom-used breathing machines do exist now to make general anesthetics usable.) This makes sense: accidentally taking a breath underwater can cause drowning. For this reason it is impossible for a dolphin to go to sleep, at least not in the sense that we do. During the night, dolphins will enter periods of decreased activity, sometimes resting on the bottom of their habitat, or swimming slowly, and coming up for air at regular intervals. Observing that the animals occasionally close one eye, rest for a time, and then close their other eye for a while, scientists suggest cetaceans may be shutting down one half of their brain at a time, leaving the other half to control swimming, breathing, etc. Because this would require both halves of a dolphin's brain to be capable of running the show, it is possible each half is a wholly functional brain in its own right. The fact that each half possesses its own blood supply may be evidence, albeit weak evidence, that this is the case. If each half of a dolphin's brain is indeed capable of independent action, they may require extra gray matter to do so, resulting in a disproportionately large brain.

In the early 80's a couple of neurological researchers came up with a very interesting theory about the function of dreams. Dr. Graeme Mitchison and Dr. Francis Crick (the same scientist who co-discovered the structure of DNA) were constructing artificial neural nets in an attempt to figure out how the human brain works. After some less than successful attempts at modeling a cluster of brain neurons, they surmised that over time bits of stored information were filed on top of one another and getting jumbled.

To test this idea, a mechanism was designed which would separate confused "memories" while the system was offline. When this worked, the scientists naturally wondered how the human brain copes with this problem. REM sleep, they reasoned, may very well prevent the human brain (and presumably the brains of other mammals as well) from overloading with information. (REM sleep is the deepest stage of the sleep process, and is the point at which we dream.) Dolphins do not experience REM sleep, which may mean another mechanism prevents their brains from overloading. It may be no coincidence that a disproportionately large brain is present in both cetaceans and spiny anteaters, the only mammals known to exist without REM sleep. It still isn't known for certain if or how cetaceans sleep. Roger Payne, in his book, Among Whales, reports incidents of snoring right whales, and there is one incident on record of what appeared to be REM sleep in a cetacean.

Cetacean brain structure suggests these animals may lack significant complex structures. The neocortex is the brain's outermost layer, and many researchers consider it largely responsible for the "superior" intelligence of mammals, including humans. Cetaceans seem to have missed this step in mammalian evolution, possessing a shallower and less complex neocortex.

While none of this proves cetaceans are less prepared for higher brain functions than some land mammals, the evidence gathered from studies of cetacean brain size and structure does not really support the concept of advanced cetacean intellect.

Amazing Animal Behavior

It is no real surprise that people are inclined to believe cetaceans are intelligent given the number of complex, clever, and in some cases "moral" behaviors they exhibit. One of these is the use of "bubble nets" by humpback whales. Seafarers had been vaguely aware of bubble netting as early as the 1920's, but it wasn't fully understood or documented until Charles Jurasz successfully observed the behavior in its entirety off the coast of Alaska in the 60's. Diving beneath a school of krill or some other prey, a hungry humpback will often exhale at regular intervals while ascending in a spiral pattern. This creates a rising cylinder of bubbles which encircles and concentrates the krill, which are engulfed by the whale from below. Humpbacks adjust the size of their bubbles according to the food being trapped, ranging from small frothy bubbles for krill to larger voluminous bubbles for big fish.

Some groups of humpbacks band together in hunting "teams", with specific individuals performing the same tasks each time. One whale leads each foray, flanked by a formation of other whales. These cooperative feeding strategies are well coordinated, and appear timed by one or more sounds made by the leader. Many theories about humpback feedings are unproven, but clearly the animals employ a high degree of specialization and cooperation.

Another astounding behavior specific to humpbacks is their propensity for singing. Almost anyone with an interest in whales has heard a recording of humpback songs at some point or another. Their music is haunting, melodious, and for all the effort by whale researchers like Roger Payne, James Darling, Peter Tyack, Phillip Clapham, and David Mattila, it is still a mystery.

Humpback songs are so called because they are composed of repeating themes and phrases, just as the songs of many birds, insects, and frogs are. Only male humpbacks sing, doing so for the most part in the warmer breeding waters of winter. Songs do occur in colder feeding waters during the summer, but these are more sporadic, and do not seem to include all elements of the "breeding season" songs. All of the humpbacks in a given ocean appear to sing the same song. Each year these songs change slightly, until after five years they are completely different. In over forty years of North Atlantic recordings, no one year's song has ever been repeated.

The most popular theories about humpback songs suppose they are mating rituals, either as a means of attracting females, deterring males, or a little of both. Some believe the songs allow the males to show off how long they can hold their breath. Others think the whales are singing about the trials and fortunes of the previous year, where food was abundant, how best to hunt, etc. What is known for certain is the songs are very complex, so complex as to permit comparison to human music in areas of rhythm, theme, note intervals, and use of percussive sounds. Even rhyme is used in humpback songs, most commonly in long, complex compositions. It would not be surprising if a whale's use of rhyme serves to ease the task of remembering phrases, as it does in human music and poetry.

Many people have been impressed with what has been termed epimeletic, or succorant, behavior in cetaceans. This refers to stories like drowning humans being carried to safety by dolphins, whales defending one of their own from whalers, and the like. Well documented cases like these aren't very common, but some do exist. On July 25th, 1976, a group of thirty pseudorcas stranded on one of the Dry Tortugas islands, about 70 miles west of Key West Florida. James Porter, a reputable marine scientist, described two particularly significant events in an Oceans article the following year. The stranding occurred along a flat, sandy beach in about a foot and a half of water during a period of low tidal activity, permitting the whales to leave the area any time they chose. Only one of the whales, the largest male of the group, appeared to be seriously ill. This individual bled slowly from its right ear, and seemed to be having difficulty, even in such shallow water, keeping its blowhole above the surface. The whale died three days later. A necropsy revealed a massive parasitic infection in its sinus cavities, a common malady amongst whales.

None of the other whales seemed to be seriously sick or injured, apart from a fair number of small wounds thought to be inflicted by cookie-cutter sharks (so named for the shape of the wounds they inflict). The whales were spread out in a wedge-shaped formation, half of the group flanking either side of the large injured whale. The shark bites were by no means life threatening, but the injuries found on the whales' backs were exposed to the sun. Despite the pain this must have caused and the inherent risk of remaining beached, the entire group resisted all "rescue" attempts and stayed with their injured comrade until he succumbed to the inevitable three days later. The whales were not hostile towards their human benefactors, and allowed suntan lotion to be applied to their skin. Yet they would become highly agitated if pushed out and away from the group, and would use their considerable strength to force their way back. Once the animals came back into physical contact with one another, they became calm once again.

How can we determine the significance of these events? Porter suggests the whale may have sought relief from the strain of remaining afloat. A parasitic infection of this kind may interfere with or disable a whale's ability to forage, and the necropsy performed turned up no food in the whale's stomach. Malnutrition and dehydration could have made the effort of surfacing to breathe unbearable for the whale. Was this the group's leader, its cohorts remaining with him out of instinct and departing only when he was obviously dead? Or does this behavior represent something more empathic, a recognition of pain and suffering followed by a resolve to comfort and protect a comrade until he recovers or dies?

Another interesting phenomenon to arise from this incident is the whales' response to Dr. Porter's approaches with snorkel gear. The first time he snorkeled into the midst of the whales, the closest broke away from the group, swam under him, lifted him nearly out of the water, and gently deposited him on higher ground before returning to its companions. Curious, Porter repeated his approach three times on each side of the massive formation with the same result. Casting aside the snorkel, Porter was able to approach in an identical manner undeterred. His suggestion that the snorkel may have sounded similar to a clogged blowhole, and the implied supposition that the whales may have been attempting to render him assistance, are speculative. Nevertheless, the incident is consistent with many undocumented reports of cetacean "aid". "Based on my own experience in this instance," writes Dr. Porter, "I would not automatically discredit such accounts of rescues by cetaceans." That certainly seems reasonable.

If the carrying off of James Porter was meant to benefit him in some way, and not to protect the dying whale, how intelligent can we consider the action? Compassion of any kind seems, to us at least, a "higher-order" process, dependent on the ability to recognize suffering in another. But in this case the whales would have been mistaken in their concern (since the scientist was in no real danger of drowning) not once but six times. Supposing they had mistaken the sound of the snorkel for a clogged blowhole; would it not have been obvious after the second incident that the "animal" (Porter) was not in distress? Considering that an even slightly impaired airway could prove so fatal to animals which face the possibility of drowning every day of their lives, it is not improbable that an instinctive response would develop over the millennia to help distressed comrades. If such were the case, it would be no miracle for an animal to generalize the behavior to include a "distressed" human, if the sound (stimulus) were similar.

Some critics of "dolphin rescue" stories cite numerous incidents of mother dolphins continuing to push deceased calves to the surface and carry them around for days, even after the corpse becomes severely decomposed. This occurs in many land-based species as well, including chimpanzees and elephants. It's difficult to decide just where instinct ends and what we call "intelligent behavior" begins. Who can say that the pilot whales' reactions to their injured comrade and to James Porter weren't just genetically preprogrammed responses? Even behavior as complex as bubble netting may have simply evolved over millions of years of evolution.

The cooperative feeding strategies of humpback whales is impressive, but not really unique. Many land mammals hunt in packs, including wolves, African wild dogs, cheetahs, and even chimpanzees. At least the wild dogs, and probably several other species, follow directional cues from a leader.

Vervet monkeys, for example, possess a limited vocabulary of alert sounds. When a vervet monkey notices a predator, it makes a corresponding sound. Each call produces an appropriate response in the other members of the group. If the call for a leopard is heard, the monkeys will flee to the trees. The eagle warning sends them scrambling for cover. These calls might be considered words, since they each represent an abstract concept. Of course, these "nouns" form the extent of vervet vocabulary, and they use them only as warnings, not to carry on conversations.

Yet "conversations" do occur in other species. Ants utilize chemical communication to tell each other who they are and whether they need food. They can warn one another about nearby enemies, or request help in harvesting a newly discovered food source. Honey bees have long been known to use movement to tell each other where nectar and potential nest sites can be found. Returning scouts will waggle their body back and forth in front of their comrades. The number of waggles the bee produces indicates how far the target area is, and the direction of the waggles indicates the desired direction of travel in relation to the sun. The bees even adjust their direction as the sun moves.

Communication isn't the only seemingly intelligent behavior in nature. Travelers in the rain forests of Central and South America are sometimes surprised to see little conveyer belts of leaves winding through the brush. Leaf-cutter ants chop relatively large chunks of foliage from trees and carry them back to their nest. The chunks are glued into place with spit, and used to grow a particular fungus, which is then consumed by the ants. The cuisine leaves much to be desired, but the farming process is very sophisticated.

Beavers are just as impressive as engineers. (I get a kick out of envisioning beavers evolving. Somewhere in beaver ancestry, a little fuzzy critter with big teeth gets weary of searching for shelter, decides he's mad as hell and isn't going to take it anymore, and gnaws down a tree.) It's hard to decide whether beavers started making dams or lodges first, but they construct both with astounding skill. Trees are cut down, disassembled, and dragged or floated to the building site. Dams are erected across streams to form ponds, which give the beavers protection from predators. The beavers know how thick to make the dams, interweaving the branches for strength, and sealing the cracks with mud and grass. They ensure the pond is deep enough to prevent it from completely freezing in the winter. These structures are often more than 100 feet in length, and can reach ten times that size. Their lodges are constructed to repel even the most curious or persistent intruders, are well insulated, and can only be entered through one of several underwater tunnels.

The Nature of Intelligence

All of these examples are impressive, and there are thousands more to choose from, but do we consider beavers, ants, bees, or even vervet monkeys intelligent? I don't think so. They are marvelously good at what they do, but their skills are highly specialized and limited to their environment. Each of these animals, and every other animal on the planet for that matter, simply evolved to be the most efficient creature in its own particular niche. But if evolution is so good at astounding us with such complex, sophisticated, yet unconscious methods of solving problems, how are we to judge whether any animal's behavior is intelligent? Any scientist worth their weight in Bunsen burners will tell you that "intelligence" is one of the most slippery, subjective terms in existence. In January of '83 Dr. Chiam Heng-Kheng, the Supervisory Psychologist of Malaysian Mensa, described intelligence at an international Mensa meeting to be "mental abilities or brightness, and has something to do with 'problem-solving abilities', 'learning capacity' or 'abstract thinking."

If even the Mensa guys are that confused about intelligence, we're in for a rough ride. I won't make matters worse by trying to list the myriad definitions I've come across during my research. My own favorite definition, and the one which makes the most intuitive sense to me, is shared by Dr. Louis Herman as well as Dr. James Gould of Princeton's Biology Department. Dr. Herman calls it "flexibility"; the ability to process information having nothing to do with a given animal's natural world. Dr. Gould describes intelligent behavior as something outside an animal's normal repertoire, something too unlikely to have resulted from evolutionary programming. In my mind, intelligent behavior requires the critical step of contextual analysis. In other words, can the animal in question compare novel situations to previous experiences, and choose to apply lessons learned to new problems?

Even if that is a valid definition, it's still near to impossible to tell when an animal is choosing an action rather than reacting "instinctively." Let's take a quick trip to the wonderful world of operant conditioning. Suppose I were to carry a big mallet around, and every time I saw my friend Rob, I yell "HAMMER TIME!" and whallop him on the head. Pretty soon even Rob gets the idea, and ducks when he hears me yell. Nothing very intelligent about Rob, right? He's simply come to associate a phrase of mine with a painful lump on the head, a result one could probably reproduce with a chicken, or whippet dog. But suppose I send another friend, Kris, to stand next to Rob with a brick and hold up a sign with the words "HAMMER TIME" printed in large, friendly letters in plain view. Rob ducks! This time it isn't such a clear-cut case of conditioning. Rob has generalized the lesson he learned concerning me, the audio signal, and a mallet and applied it to Kris, a written phrase, and a brick without any assistance or instruction. That's fairly impressive. (Well, for Rob it is.) But what if I had simply walked up to Rob with the mallet and had Kris yell the magic phrase? If Rob ducks, is he thinking, or simply reacting like a Pavlovian dog?

This is a poor example to use for discussing cetacean intelligence, partly because it involves negative reinforcement, but mostly because humans have a natural advantage in generalizing between a spoken phrase and a written one. But it does illustrate the difficulty in targeting an exact point at which an animal's behavior becomes the result of conscious analytical thinking. Animal trainers have known for years that cetaceans are capable of learning new and often highly complex behaviors. This is not a basis for judging intelligence, since any animal can be trained to some degree. Many animals, including dogs, horses, and pigs, are just as capable of learning complex tasks as cetaceans. It may be easier to judge an animal's thinking ability when it is presented with challenges completely alien to it.

Can Any Dogs Learn New Tricks?

Animal intelligence studies have challenged creatures with a number of foreign concepts and puzzles. Many mammals can learn artificial languages. Chimpanzees and gorillas have been taught constructed symbolic languages with greater success than cetaceans. Konzi, a Benobo (pygmy chimpanzee) under the care of researcher Sue Savage-Rumbaugh at Georgia State University, has learned more than a hundred words, each represented by a symbol. He can respond to questions and requests, and make requests of his own (which usually involve tickling and biting.) He can generalize basic rules of syntax and grammar to respond to novel requests correctly on the first attempt. Koko, a lowland gorilla at a The Gorilla Foundation in Woodside, CA, has been learning sign language from Dr. Penny Patterson since 1972, and is now supposed to have a working vocabulary of more than 500 words. Sally Boysen, of Ohio State University, is making progress teaching hand signals for objects to Vietnamese potbellied pigs. Alex the parrot and Irene Pepperberg at the University of Arizona have proven that even some birds can learn to verbally identify shapes, colors, materials and quantity.

Some of the most interesting work in mapping out dolphin intelligence occurs at Kewalo Basin Marine Mammal Laboratory in Hawaii. The organization's director, Dr. Louis Herman, studies dolphins' ability to understand artificial languages. Two separate languages, one made up of hand signals and another of auditory signals, have been used to teach dolphins to understand and react to somewhat complex requests from humans. The purpose of these studies is not to converse with the dolphins, but rather to better understand how they process information, adapt to new situations, and solve problems.

The Lab's dolphins have learned over 50 words which may be arranged in sentences of up to five words forming a request. The dolphins can tell the difference between "take the ball to the surfboard" and "take the surfboard to the ball." Similar to Konzi and Koko, they have taken this basic principle of syntax and extrapolated its use to include other objects and actions, usually responding correctly on the first attempt. They can put things on top of other things, swim under, over, and around objects, etc.

The Lab's dolphins aren't limited to such unsophisticated tasks, but have proven to possess a limited capacity for abstract thought. They can apply the concept of "ball" to any spherical object with which they are presented. They can confirm the presence or absence of an object in their pool. They can remember new sounds and objects for several minutes after they encounter them. They can accept visual cues from television images of their trainers, images of their trainers' hands, or even white spots of light tracing the hands' movements. In fact, the very first time the dolphins were ever presented with a television image of a trainer issuing an instruction, the animals executed the requested behaviors flawlessly.

Clearly, dolphins possess at least some capacity for taking learned sets of rules or principles and applying them to novel situations. Is that evidence of significant brain power? Does the dolphin equivalent of "hmm...let's see now..." ever pass through their minds?

Some researchers in animal intelligence believe self-awareness is the key to answering such questions. At least two studies have been conducted on dolphins' ability to recognize mirror images of themselves. Similar studies have been carried out with primates, the results of which strongly suggest that many of the great apes understand the nature of their own reflection. The 1994 account of the Earthwatch mirror study by Kenneth Martin and Suchi Psarakos, printed in the book Self Awareness in Animals and Humans, makes a pretty compelling case for self-awareness in dolphins. At the very least, it seems obvious that the dolphins knew they were looking at themselves.

Where WILL the debate end?

There are many more points of view on dolphin and whale intelligence than those I've expressed here. If you have an interest in resolving the issue for yourself, I strongly suggest taking a look at the reading list on page 353, and get started. Try to get as much personal experience with the animals as possible. A picture is worth a thousand words, but a little first-hand exposure is worth ten times as much.

What's that?

What do I think? Oh, boy. I suppose I should leap at the opportunity to express myself before the masses, but you can't help but wonder, at times like these, if you'll one day be shown to have gone on record with the wrong opinion. Ah, well, here it is:

If you're looking for a way to talk to dolphins like Dr. Doolittle, give it up. There has never been any evidence to suggest that dolphins or whales think the way we do, or use a language even rudimentally similar to our own. If on the other hand, if you're trying to build a case that cetaceans are fairly intelligent animals deserving of respect and consideration, you can easily do so using irrefutable scientific evidence. It can be said with confidence that cetaceans are very adaptable and social creatures which exhibit complex cooperative behavior and even a certain degree of self-awareness.

Part Two

Cetacean Careers

- ►Display Facilities
- ►Research Organizations
- ►Conservation Groups

4
Display Facilities

So, you want to be a dolphin trainer, eh? It's a tough game, but if you're following your dreams, and you've got what it takes, you just might make it. Read on, intrepid ones, and I'll tell you what you need to know.

What is a dolphin trainer?

Thirty-six organizations house dolphins and whales in the United States and Canada. All together they run forty-two locations, each staffed by people responsible for the care and training of the animals. Some are called dolphin trainers. Others are known as animal care specialists, naturalists, biologists, mammologists, or technicians, but they're all roses, if you take my meaning. For simplicity, we'll just call them all "dolphin trainers".

Dolphin trainers might work exclusively with dolphins, or might be responsible for manatees, seals, walrus, polar bears, otters, sea lions, or any number of mammals, fish, birds, or invertebrates. A dolphin trainer's responsibilities vary greatly from facility to facility. Here they might do public speaking, there a bit of water quality analysis, or even research projects. Some work with stranded and abandoned animals, or lecture school groups. Some work indoors in cold water, while others spend each day outdoors in the sun.

As a rule, dolphin trainers aren't highly paid. In 1995, the lowest-paid, full-time, entry-level trainers made less than six dollars per hour, while the highest rate was more than twice as much. The majority made about eight dollars per hour. Almost all dolphin trainers get health insurance of some kind, and the larger institutions often include things like tuition reimbursement, 401(k) plans, and other perks. Nobody gets rich being a dolphin trainer, but it can provide an adequate living.

Advancement is limited in animal care, but some trainers become managers, curators, or directors. Others branch out into research, conservation, or other areas of animal care like elephants, birds, fish, or reptiles.

What is being a dolphin trainer like?

Brace yourself: here come the cold, hard facts. Being a dolphin trainer is a rough job, probably one of the most demanding jobs you could have. You'd probably start each day around seven AM, and spend an hour or two sorting through hundreds, or even thousands of thawed fish. Each fish has to be checked for breaks in the skin and placed in a bin according to its size. Once screened and binned, a specific daily amount of fish is weighed out for each animal, and divided into portions for each feeding or training session.

Why get married when you can wake up to this each morning?

The whole process ensures you start each day full of scales and fish blood, and smelling just dreamy. Regulations in the U.S. require the "kitchens" where this is done to be completely cleaned up after each use. This means you spend more time cleaning sinks, shelves, buckets, bins, and floors than you do anything else. The constant bombardment of cleansers, fish scales, cold salt water, and hot tap water will turn the skin on your hands into a nice lizard-like surface, and if you're in an area where it snows, your fingers will simply crack open and bleed at random intervals in the winter.

You know you're working as a part of a team because you spend more time with your co-workers than with your family. If one of you gets sick, you all get sick, and since you are usually walking around in soaking wetsuits, it's happening all the time. Speaking of wetsuits, once you're done sorting fish and cleaning the kitchen it's off to the locker room to get changed for that 8:30 whale feed. By 8:50 the feed's done and you're washing the buckets you used. You grab the food for the otters at 9, pausing just long enough to write down totals for the food you sorted and behavioral notes from the whale feed. Otter session ends: back to the kitchen (wash wash wash). It's only 9:25, so you help the volunteers for a few minutes cutting squid for the otters. 9:40, time to go feed the penguins.

Brr, it's 37 degrees in there, and your feet are still wet from the other animal feeds. Still, the little critters are pretty cute. Penguins eat, bite, and crap on you. Little bastards. Back to the kitchen (scrub scrub scrub). 10:05, just enough time to warm up in the office and write down notes on the otters and penguins. Off to narrate the first dolphin presentation. You switch yourself on automatic, sit back, and let your mouth do the driving. "Dolphins are warm-blooded mammals..." blah blah blah. You do the feeding, cleaning, and recording for three more animal sessions, and stagger into the office for a well deserved lunch. Just as that first bite of your turkey sandwich is headed for your mouth, a call comes in from security; you have to go retrieve a pair of sunglasses from the bottom of the dolphin habitat. Damn. There must be some guy who comes each day and waits for lunch to drop stuff in. Smile, the day's only half over.

That's actually a pretty typical morning at the Shedd Aquarium, and the experience isn't all that different at other facilities. Dolphin trainers have little spare time, and not a whole lot of energy when they get off work.

So why on Earth would anyone subject themselves to such a harsh life? The answer lies in our fascination with dolphins and whales. They're absolutely incredible. Even viewed through a telescope or TV, cetaceans awaken something wondrous inside us. Up close, the effect is increased tenfold. Caring for such animals, teaching them, and playing with them gives you feelings you'll never find elsewhere. Feelings of accomplishment, of awe and respect.

You'll learn their personalities, temperaments and quirks. You'll know the games they like to play, what their favorite toys are, and whether they're "morning" dolphins or night owls. They'll figure you out too, and just when you think you understand them, they'll surprise you with something you'd never have predicted. Every so often they even remind you that humans can be trained as well.

These sorts of people don't mind the long hours or the harsh physical demands. For them, the time they spend with the animals and the relationships they build with them are satisfaction enough.

Are you such a person?

The only way to truly find out if you are is to get a little first-hand experience. Get out there and volunteer some of your time. There are more than two dozen zoos and aquariums in this book which have volunteer programs. If you aren't near any of them, don't panic. Almost any kind of animal experience will tell you whether this is your thing. Go help out at a horse stable, veterinary clinic, animal shelter, or even a pet store. Feeding and cleaning are pretty similar processes for any animal facility. After a few weeks, or a couple of months, you'll start to figure out whether you're cut out for it.

Penguins can be a real treat to work with.

How can I get to be a dolphin trainer?

Brace yourself again, 'cause here come some more hard truths. As tough as being a dolphin trainer can be, getting a job as one is even harder. There are only about 400 of them in North America. By contrast, the first print run of this book was 2,000 copies, and it nearly sold out before rolling off the presses. A 1995 poll of seven marine mammal display facilities indicates 40,000 people requested information about becoming a dolphin trainer in 1994, and the number of total inquiries industry-wide is guaranteed to be many times that figure. There is fierce competition for every single dolphin trainer position which opens up in North America. The only way to get one is to be in the top 1% of the applicants, and to be in the right place at the right time. Freaking out yet? Don't worry. If you're really cut out for dolphin training, you'll make it. If you're not cut out for it, it won't be long before you realize the fact, and you'll be on your way to figuring out what you really want to do with your life. Don't forget, there are a lot of other kinds of cetacean groups in this book as well.

Now, let's deal with getting you into that top percentage. You only have control over part of what makes you a good applicant. This is the part which includes your education, experience, and attitude. The other part, your personality and compatibility with the staff you're trying to join, isn't something you can change. Because the management at zoos, aquariums, and marine parks have so many applicants for their jobs, they can afford to be very, very picky.

Some people may simply never get hired, regardless of their credentials or effort, because the right chemistry with the staff just isn't there. Teamwork is an absolutely essential component of an animal care program. Hiring a qualified but incompatible applicant would not be a decision made with the animals' best interests at heart. Hopefully, in such a case the management would be honest and forthcoming with this information, since anything less is unfair to everyone involved. I don't want to overstate this point, but it's important for you to be prepared for the possibility that some doors will remain closed to you despite your best efforts. If you run into this situation, it may be time to try another facility someplace else.

All right, let's see if we can't be a bit less grim for this next section. One of the most common questions I got asked at the Shedd Aquarium was "where should I go to school?" If you read the Introduction, you might have figured out that Marine Biology programs have very little to do with cetaceans. In fact, almost no dolphin trainers have degrees in Marine Biology. Far more common degrees are Biology, Zoology, and Psychology. Even the Biology and Zoology programs of most colleges have nothing to do with marine mammals, but they will give you the general background in the biological sciences that most zoos and aquariums look for.

Most dolphin and whale facilities accept a Psychology degree just as readily. Much of what dolphin trainers do is operant conditioning, a topic covered in great detail by most psychological programs. Strange as it may seem, an education in Psychology will give a dolphin trainer more practical job skills than a degree in the biological sciences.

Don't despair if you've already graduated with a different degree. Many facilities will hire someone based on their other qualifications, provided they have a degree of some kind. You may not be quite as strong of a candidate with an English, Engineering, or Communications degree, but such people have become dolphin trainers in the past.

Some people have managed to get a job with a two-year degree, or even no degree at all. Don't get your hopes up for this. Remember that your goal is to make yourself the very best candidate you can, and part of that is to get a four-year education. If, for some reason, you are presented with an exceptional opportunity to gain full-time experience in marine mammal care, then it might be advisable to forgo the college degree, at least for the time being. Otherwise, find yourself a college you like which is close enough to a marine mammal facility of some kind, so you can volunteer or intern in your spare time. (More on this in a bit.)

There is also Moorpark College to consider. Moorpark offers the only college program in North America which teaches you to care for and train exotic animals, including marine mammals. Moorpark has developed relationships with the marine mammal display field, and each student spends two weeks working at a host facility prior to graduation. It's important to note that Moorpark is a two-year college. Graduating from Moorpark may improve your chances at getting a job in animal care, but it might also limit your options. Some marine mammal facilities won't accept anyone without a four-year degree no matter what. Career advancement within the field requires at least a Bachelor's degree, and often a Master's as well. See page 350 for more information on Moorpark College.

Now let's talk about experience. Volunteering at an animal facility isn't just a way to get a feel for the business. It's also a good first step in building your resume. Don't expect to get a job as a dolphin trainer without animal care experience. It's been done, but very, very rarely. Volunteering at a zoo or aquarium is going to be one of your most valuable tools in pursuing a job. To make the most effective use of volunteering, do so at a facility you'd like to work at while enrolled in college. This way, you're accomplishing four things at once:

1. You're getting your education. In fact, you'll be in a position to integrate what you do as a volunteer into your studies. Seek out help from a professor with similar interests, and see if the two of you can collaborate on informal projects, or even arrange for independent study credit.

2. You're getting to know what the job is like while your options are still open. If you decide you'd rather go into cetacean research, or hairstyling, you can adjust your studies accordingly.

3. You're getting experience in animal care. Even if you never come into direct contact with the animals, zoos and aquariums look very favorably on consistent, dedicated volunteer work. They'll also know you understand the rigors of working in their field.

4. You're building a relationship with the staff. This is a huge benefit. Getting your foot in the door is one of the hardest parts of becoming a dolphin trainer. Volunteering gives you a chance to get to know them, and they get a chance to get to know you. This is the part where the aforementioned compatibility issue comes into play. If you are compatible with the staff, volunteering with them will do wonders for your chances. Of course, if you aren't, it won't.

Volunteers clean and prepare food more than anything else. Some facilities allow animal contact to a limited degree, and a very few use volunteers to work directly with the animals. The majority reward their volunteers with occasional play sessions, where touching the animals is permitted under staff supervision.

When you're volunteering, don't forget you have to be in that top 1%. Try to excel at everything you do. Don't knock yourself out trying to please everyone, because you'll come off as a suck up. Just do the best job you possibly can, and go the extra mile when appropriate.

Internships are another terrific way for students to get a taste for the job. Internships can last one day, or an entire semester, depending on the facility. Internships can involve no animal contact, or may treat the student as an entry-level trainer. Most internships are only open to college students, though a few are open to anyone. Check the facility descriptions to find out what the various internships offer.

If you're already in school, and there isn't a zoo, aquarium, or marine park within reach, find some other place to volunteer. (See the last paragraph under "What is being a dolphin trainer like?") There are lots of other kinds of experience that will help you as well. Speech classes, rhetoric clubs, or any job experience in public speaking is very helpful. Most dolphin trainers spend a lot of time answering questions and giving presentations.

Get yourself SCUBA certified as soon as you can, and get a little diving experience once you do. Contact local dive shops and compare some prices on openwater certification courses. The courses are affiliated with various certifying organizations, the most reputable of which are NAUI, PADI, and SSI. You should also get a CPR certification, and take a first aid course, both of which will have to be renewed every year or two. These aren't usually requirements, but every little bit helps.

You should start learning about dolphins and whales, the display industry, conservation, animal care, and training techniques as soon as possible. There is an excellent reading list on page 355. If you can find classes on dolphins and whales, take them. Many organizations listed in this book offer one-day courses on marine mammals, and the Schools list on page 350 details most college courses on dolphins and whales in the U.S.

As a rule of thumb (what DOES that phrase mean?) any kind of involvement with the organizations listed in this book will look good on your resume. Stranding response, research, and conservation work would be particularly helpful. A few possible exceptions are ORCALAB, the Cousteau Society, Earth Island Institute, the Great Whales Foundation, the International Wildlife Coalition, Save the Whales, Dolphin Alliance, and the International Dolphin Project because of their opposition to the display industry.

Last, but not least: the 'TUDE. Attitude can go a long way toward improving that compatibility issue. You can't change your personality, but you can train yourself to give 110% to everything you do. Dedication, persistence, heart, chutzpah; these are necessary traits for those wishing to enter this field. If you can't hang tough, go find something else to do.

That's everything you need to know to turn yourself into a good applicant. As far as being in the right place at the right time goes, there actually is a bit you can do to help. On page 251 is a description of the International Marine Animal Trainers' Association. Get yourself a membership and keep abreast of changes at marine mammal facilities. When a major expansion occurs at a zoo or aquarium, you can be pretty sure they'll be hiring new people soon. Watch for pregnancies, too. If one or more marine mammals become pregnant, a little math will tell you when the staff might need to hire on temporary help.

You might want to consider getting a membership to one of the anti-display organizations listed two paragraphs above. They tend to know faster than anyone else when a new display facility is being built, or when new animals are going to be acquired, which are good indications of impending job openings.

Go to the conferences for the professional associations listed in chapter 8. Learn from the presentations, and don't be afraid to schmooze if the opportunity presents itself. Try to meet people involved in facilities and projects you're interested in. Ask them about what they do, and tell them what your interests are. Even if nothing remarkable results from the conversation, you'll learn a lot, and get to know people.

One last piece of advice: THERE ARE NO RULES. What I've given you here are guidelines, not laws. For every recommendation I've made, there are a handful of people who went the other way and succeeded. We all found our way into this business by our own path, and no two stories are exactly alike. Use that big blob of gray goo in your skull, and come up with ideas of your own. Don't be afraid to take chances. Find out as much as you can about where you'd like to go, and use the information to your advantage. Never stop learning, and if you're not sure about something, ASK. Remember, these people are looking for exceptional candidates. BE exceptional.

Here's a cheery factoid: Fewer than four millionths of a percent of Americans get to be dolphin trainers.

Display Facility Descriptions

Okay, enough chit-chat. What you're about to read are descriptions of all the thirty-six groups which house cetaceans in artificial or enclosed habitats. Each entry is divided into four sections. "Overview" gives you a general idea of what the place is all about. "Employment, Intern, and Volunteer Opportunities" you can probably figure out by yourself. "Research" describes what research programs, if any, are currently being done with that facility's cetaceans. "Other Programs" is a quick run-through of educational, conservation, or other useful programs offered by the organization. Don't forget to check the Omega website every few months for updates, since this field changes rapidly. (www.omegaland.com)

Good luck!

Aquarium For Wildlife Conservation

Surf Ave. & West 8th St.
Brooklyn, NY 11224
Phone: (718) 265-3405
Fax: (718) 265-3420
http://www.nyaquarium.com

In 1895 a group of prominent New Yorkers banded together "to establish and maintain in said city a zoological garden for the purpose of encouraging and advancing the study of zoology, original researches in the same and kindred subjects, and of furnishing instruction and recreation for the people." They accomplished this in 1899 when the 265-acre Bronx Zoo opened in November of that year. This gave birth to the Wildlife Conservation Society which now owns and operates the Aquarium for Wildlife Conservation.

The Society recently celebrated its Centennial birthday. Over the past century it helped establish more than a hundred wildlife parks and reserves around the world, and conducts about 300 conservation field projects in over 50 countries worldwide. It is largely responsible for the tremendous shift toward education and environmentalism that has swept through the Zoo and Aquarium industry over the last sixty years. The Society currently operates five zoological parks in New York City - the Bronx Zoo/Wildlife Conservation Park, the Aquarium for Wildlife Conservation in Brooklyn, the Central Park Wildlife Center, the Queens Wildlife Center, and the Prospect Park Wildlife Center.

Overview

When the then-named "New York Aquarium" opened in Brooklyn 1957, it inherited the legacy of its predecessor, the aquarium at Castle Clinton in Manhattan. The Society assumed responsibility for the old aquarium in 1902 at the request of the New York City Parks Department, which believed the Society would be better suited to implement an educational mission for the park. Though it had been showing its age in 1940 (the year it closed) the original New York Aquarium was an impressive structure. Its main atrium, decked out in pillars, archways, and sculpture, featured four massive central pools, 97 large wall tanks, and circulated over 300,000 gallons of water per day. Unfortunately the entire building was demolished to make way for a proposed bridge in the early forties.

(The bridge, by the way, never did get built.) The aquarium's personnel, animals, and equipment were temporarily relocated to the Bronx Zoo with the promise from the city's Commissioner of Parks that a new $10 million aquarium would be built to accommodate them. The money never surfaced, however, and in 1957 a much more modest facility opened in Coney Island, Brooklyn. Because the new aquarium's funding has never been quite commensurate with its status or accomplishments, it has been a work in progress ever since, with new add-on exhibits, habitats, and improvements appearing every few years or so. Consequently, while it has done more to advance animal husbandry, research, and education than its predecessor, it lacks some of the elder facility's aesthetic beauty.

Today, the Aquarium for Wildlife Conservation houses nearly 4,000 animals representing over 300 species, including beluga whales, bottlenose dolphins, seals, sea lions, sea otters, and walrus. The marine mammals are cared for by two distinct groups; the Training Department, which handles the cetaceans and some of the pinnipeds, and the Sea Cliffs staff, which is responsible for the otters, walrus, penguins, and the majority of the pinnipeds. As you might guess, conservation and education are the watchwords of the Aquarium, and their exhibitry and public presentations reflect this. The Aquarium is still in a state of expansion, with upcoming plans for new exhibits and improvements to existing habitats.

Employment, Intern, and Volunteer Opportunities

New York is a hard place to live. It's crowded, noisy, and dirty, and it isn't particularly safe. The cost of living alone would drop a Nebraskan's jaw right on his feet. Nevertheless, for those determined to live there, New York can provide opportunities found nowhere else.

There are about a half-dozen full-time staff in the Aquarium for Wildlife Conservation's training department, and about nine in the Sea Cliffs department. They care for the whales, dolphins, and pinnipeds in the Aquarium's AquaTheater and beluga whale habitats, as well as the Sea Cliff's outdoor naturalistic habitats. The Aquarium's trainers narrate sessions, operate the audio booth, perform water quality analysis, and any of a host of other tasks. A college degree is not necessary, but is always helpful. An interesting side note is the Society's policy of reimbursing employees 100% for the cost of passing-grade college classes which relate to their job. This is almost unprecedented in the field, and more than one employee has used this policy to help pay for a degree. Positions tend to open up once every two or three years, and are usually awarded to prior or present seasonals, interns, and volunteers.

Seasonal employees are taken on for a number of months, fulfill much the same functions as full-time trainers, and may come back as many times as they are accepted. They are paid, though at a substantially lower rate. The number of seasonal employees at the Aquarium fluctuates, though generally does not exceed three at one time.

Interns come most often from Southampton University, which has developed a relationship with the Aquarium, though other students may apply as well. An internship must qualify for college credit at the student's home campus. Less emphasis is placed on having interns complete a singular project than on giving them hands-on experience with marine mammal care. These internships are unpaid, and usually last one semester.

Volunteers serve much the same function as interns, spending the majority of their time cleaning, preparing food, etc. Many seasonal position applicants have come from the volunteer ranks, and it's a good place to start if you'd like to work at the Aquarium.

Both Interns and volunteers have an chance to gain tremendous experience at the Aquarium, as their responsibilities and opportunities increase directly in proportion to the amount of enthusiasm and competency they demonstrate. (Even more so here than at most other facilities.)

Research

Being an extremely active participant in global environmental and conservation research, the Wildlife Conservation Society is in an excellent position to conduct in-house research. This gives Aquarium employees access to some of the most accomplished conservation researchers in the world. The Society gives out a limited number of internal employee research grants as well. Previous studies have focused on the visual acuity of Belugas and self-recognition in cetaceans. New proposals are submitted all the time, and those which are compatible with the Aquarium's mission and structure are given full consideration.

Other Programs

The Society's education programs reach almost two million schoolchildren each year. The Aquarium offers dozens of classes and workshops for students of all ages, including several introductory courses in marine mammology. Contact the Education

Department for full details. Annual family memberships are available for $58 and include unlimited admission to all five zoological parks and a one-year subscription to the Society's highly-acclaimed Wildlife Conservation magazine.

Cedar Point

P.O. Box 5006
Sandusky, OH 44871-8006
Phone: (419) 627-2350
Fax: (419) 627-2200
http://www.cedarpoint.com

"Sandusky?! Where the heck is that?" If you've never heard of Cedar Point before, you'll probably be surprised to hear there's more to this little town than the Goofy Golf on Route 250. Located on the southern shore of Lake Erie, Sandusky harbors a number of prominent facilities, including NASA's Lewis Plum Brook Station, which tested the landing gear of the recent Mars Pathfinder probe, Coon's Candies, reportedly the manufacturer of the toffee with the highest butter content on the planet for over 75 years, and Cedar Point amusement park.

Overview

At 364 acres and featuring 59 rides including 12 roller coasters, Cedar Point is arguably the largest single amusement ride park in the world. The resort features two hotels, one of the largest marinas on the Great Lakes, performances by musicians and comedians, a water park, two miniature golf courses, and a go-kart raceway. Cedar Point employs a year-round staff of nearly 300 people, and hires on an additional 4,000 seasonal workers each summer. The park welcomes more than three million visitors each year.

Cedar Point's true charm lays not in its sheer size, but in the park's long and rich history. The park officially opened in 1870, making it one of the oldest amusement parks in the world. Between the late 1800's and the time of the Great Depression, Cedar Point achieved national fame and recognition as one of the country's most popular tourist spots. The stately Grand Pavilion, built In 1888, sports a huge dining room, bowling alleys, officers' quarters, private dining facilities, a bandstand and live stage, a dance floor, an auditorium, a beer garden, and public docks.

The older of the park's on-site hotels, the 600 room Hotel Breakers, opened in 1905. In its time the Breakers enjoyed fame and notoriety to rival even San Diego's Hotel Del Coranado and New York's Waldorf-Astoria. In the years preceding the depression it hosted such celebrities as Annie Oakley, Abbott and Costello, author Theodore Dreiser, Sam Warner (of Warner Brothers), and composer John Philip Sousa. Six U.S. presidents also stayed at the

Hotel Breakers, including Franklin D. Roosevelt and Dwight D. Eisenhower. Metropolitan Opera stars were known for giving impromptu concerts from the balconies of the hotel's five-story rotunda, and Notre Dame quarterback Knute Rockne perfected football's forward pass on the hotel's strip of Lake Erie beach in 1913. The hotel still features chandeliers and stained-glass windows designed by the Tiffany company in New York City and original wicker furniture from Austria.

Cedar Point survived the Depression, (barely) and after sixty years of restructuring and growth bears little resemblance to its former self. 1980 saw the opening of Oceana, the park's marine mammal complex. Oceana houses two bottlenose dolphins and two sea lions in a stadium-like habitat overlooking Lake Erie. The dolphin and sea lion shows are entertaining, very well executed, and educational. The complex includes the Oceana aquarium which exhibits sharks, tropical fish, living corals, and poison dart frogs.

Employment, Intern, and Volunteer Opportunities

There are only two trainers at Oceana, (the same two since the habitat's inception in 1980) so don't get your hopes up for a permanent job. There are no volunteer positions, but an internship program is being considered.

The good news is, Oceana brings on seasonal help in the summer. Three animal care assistants help with food prep and cleaning, and get involved with the animal sessions and feeding. Although competition is heavy for these positions, (people apply from all over the country) they represent an excellent opportunity for aspiring trainers to get hands-on experience with marine mammals. Oceana has taken individuals with no prior marine mammal experience, but there are a lot of applicants with an oceanarium internship or the like under their belt, so the more animal experience you show the better. Applicants are usually college students or graduates.

There are also five seasonal usher positions available, which is a good way to get involved and make yourself known to the staff if you can't snag the animal care position. Ushers need good communication skills, as their job is to interact with the public and answer questions. Oceana also takes on an entry-level audio technician each summer to run the sound booth for the shows.

All seasonal positions pay just above minimum wage, though a retroactive pay increase of about fifty cents/hr. is given to workers who complete more than 400 hours of service, and a one dollar

(Photo by Dan Feicht and provided by Cedar Point)

increase/hr. is given for hours worked beyond 400. Most seasonals from out of town cannot find their own housing, since Sandusky is a resort town and fills up quickly in the summer. Provided they live at least 35 miles away and are over 17 years of age, seasonals may take advantage of the on-site dormitory-style housing. The dorms are not glamorous, but are reasonably priced, and you can bet it will be a memorable experience. The dorms house up to 3,500 workers each summer; nearly all of the seasonal staff. Applications should be submitted no later than January for the following summer. (Earlier is better.) Positions run roughly from early April through October.

Research

Oceana's staff and time constraints make most research projects infeasible, but proposals for observational projects may be considered.

Chicago Zoological Society

Brookfield Zoo
Brookfield, IL 60513
Phone: (708) 485-0263
Fax: (708) 485-3532

Overview

The Chicago Zoological Society is based in Brookfield, Illinois, where it operates the Brookfield Zoo. (The zoo in the City of Chicago is named Lincoln Park Zoo, and is a completely separate organization.)

Great flowing, towering, menacingly huge gobs of cash is what you generally get when you cross two families like the McCormicks and the Rockerfellers. This and a healthy appreciation for animals enabled Mrs. Edith Rockerfeller-McCormick in 1920 to donate over eighty acres of Cook County, Illinois to the Forest Preserve District for the express purpose of building a Zoological Park for "the promotion of zoology" and "zoological research." It very nearly failed to happen. With funding scarce, opponents lobbied to prevent the use of local taxes for its construction, which, ran smack dab into the Great Depression. You can't keep a good zoo down, though, and through the faith and determination of some of Chicago's bigger wigs, Brookfield became one of the best. It was on the cutting edge of the then new trend of naturalistic habitats in zoos, and in the last sixty years has repeatedly broken new ground in many areas, including exhibition techniques.

In 1987 the zoo unveiled the successor to its first dolphin habitat, which was an aging, tower-like, George Jetson-looking thing with poor lighting and insufficient space for the growing collection of animals. The Seven Seas Panorama, as the zoo's new dolphin habitat was named, offers spacious accommodations for both its residents and visitors, sports a nifty skylight roof, and incorporates over a dozen different plant species, including more than 450 real trees to complete the facility's Caribbean theme. (Alas, my suggestion of adding a steel-drum band and banana chips was largely ignored.) The shows are fairly theatrical, but take plenty of time to incorporate environmental awareness. Between shows the amphitheater is closed, and the dolphins may be viewed through the habitat's many underwater windows. The area to the north of the Panorama features an outdoor Pacific Northwest seascape exhibit housing walrus, seals, and sea lions.

All this occupies just a small portion of the zoo's 216 acres, which houses about 2,500 animals. It's one of the biggest zoos in the country with some very impressive exhibitry. Tropic World, the zoo's main primate exhibit, is a must-see for even a casual animal fan.

Employment, Intern, and Volunteer Opportunities

The key to pursuing an entry-level full-time job at Brookfield Zoo's Seven Seas Panorama is to start out as a "seasonal" and work your way into a permanent position. Although some trainers have been hired without seasonal experience, most of them had been trainers elsewhere. There's no limit on how many seasons you can work, and in fact, some people have held many seasonal positions back-to-back. Even seasonal jobs are very difficult to get, though, and there's no sense in applying if you aren't at least in college and have some type of experience working with animals. Many applicants have finished a degree in Animal Care Science, Biology, or Zoology, and have experience working with marine mammals. (Many of these come from the nearby Shedd Aquarium, or were Seven Seas interns.)

Seven Seas usually employs about two seasonals at a time. Not all of them are looking to become dolphin trainers, but nearly all are trying to get a zookeeper position somewhere, so competition is fierce. Nevertheless, it seemed to me that the staff makes an effort to maintain a balance between bringing on new faces and allowing familiar ones to return. One technique which may be overlooked by some applicants is to obtain a seasonal position in another part of the zoo and build experience there. This will not only build up your qualifications, but it will put you in a position to get to know the staff. Seasonals do much of the same work as the permanent trainers, and they usually wind up working with animals on their own, though they aren't assigned behaviors to train. Like permanent staff members, seasonals may be called upon to narrate the dolphin shows, so communication skills are a plus.

Brookfield Zoo provides learning opportunities in animal management techniques through its Intern Program. Applicants must have completed at least two years of college, have maintained an overall GPA of 2.5, and have a serious interest in an animal or conservation-related career. Selected applicants may request placement in the Seven Seas Panorama, as well as in many other areas of the zoo. Interns at Seven Seas do not work directly with the animals, but gain valuable exposure to management and training techniques by observing trainers in sessions with the pinnipeds and the dolphins.

Internships in the Seven Seas Panorama are available year-round, last about twelve weeks, and are full-time and unpaid. The summer term (from May through August) tends to be the most competitive. Prospective applicants can contact the Intern Program Coordinator ate ext. 459 for more information and an application.

Research

The Society funds many research efforts in remote locations, including the ongoing wild dolphin population studies in Sarasota, Florida conducted by Dr. Randy Wells.

The Chicago Zoological Society has designed its two cetacean facilities to compliment one another by designating Seven seas as its vehicle for public education and entertainment, and the Hawk's Cay facility as its center for breeding and research. Consequently, very little research is conducted at Seven Seas Panorama, though Amy Samuels (whose recent study of swim-with-dolphin programs in the US was invaluable during the writing of this book's chapter on the subject) is actively involved in the analysis of behavioral observations recorded by the staff.

Other Programs

Like most big zoos, Brookfield offers a slew of educational and children's programs on a wide variety of subjects, as well as an assortment of outings, tours, and expeditions. Recent cetacean-related items have included trips to New England and Florida, tours of the Seven Seas facility, and lectures by Dr. Randy Wells. Volunteers are needed as educational docents, fund raisers, and office assistants. Zoo memberships are available as well. Contact the Education department at extension 361, the volunteer department at 852, and the membership office at 341.

Clearwater Marine Aquarium

249 Windward Passage
Clearwater, FL 33767
Phone: (813) 441-1790
Fax: (813) 442-9466
http://www.flaoutdoors.com/wildlife/marine.htm

Clearwater is a large town on the rather quiet west coast of Florida, just west of Tampa. Sitting on an easy-to-reach stretch of island amid fairly expensive condos and hotels is Clearwater's #1 defender of wildlife, the Clearwater Marine Aquarium.

Overview

The CMA is an active member of the Southeast Marine Mammal Stranding Network, and handles all strandings from the Tampa area north to the panhandle. Its primary mission is marine mammal and turtle rescue and rehabilitation, though is also acts as an educational clearing house on a huge range of marine topics.

The Aquarium has a few permanent residents, including two river otters and an Atlantic bottlenose dolphin named Sunset Sam, which is why the Aquarium is listed here. All of these were stranded or abandoned animals which are unsuitable for release. In addition to scheduled narrations at the dolphin and otter habitats, the center features a mangrove exhibit, a sea turtle rehabilitation tank, a stingray pool, and an assortment of educational displays.

The CMA plays a vital role in fighting habitat destruction and preserving valuable ecosystems in southern Florida. Many of Florida's residents are still unaware of how important estuaries, sea grasses, and mangroves are to Florida's wildlife. By reaching out to its thousands of annual visitors, the Aquarium has a very real and positive effect on Florida's environmental problems.

Employment, Intern, and Volunteer Opportunities

There are only six paid employees in the Animal Care Department, and staff turnover is very rare, making Clearwater one of last on the list of prospective employers. Nevertheless, the CMA offers some very good opportunities to build experience in marine animal care.

Volunteers are a big part of the Aquarium. Volunteers may choose to help out with fund raising, education, displays, or a number of other support areas. Volunteers are requested to commit to one day each week. The Animal Care Department also makes use of volunteers, who are invited to apply after a minimum of three months' service in another area. Animal Care volunteers maintain the sea turtle and stingray tanks, narrate sessions with the otters and dolphin, and occasionally assist staff members in training sessions.

Volunteers are used extensively by the Aquarium's stranding response team. About 25-30 volunteers, all of whom have extensive experience at the center and have demonstrated a high degree of competency, have been trained to respond to strandings, and are on call along with the staff members.

Animal Care internships are available for college students. Internships are unpaid and typically run for the duration of the student's current academic semester, though this period is flexible and can accommodate other schedules as well. Interns can expect to spend about 32 hours per week assisting with food prep, cleaning, record-keeping, etc.

Full-term internships usually involve completion of a research project, though shorter internships may involve training an assigned behavior with the dolphin. Here again the program is flexible, and the intern and the staff work out actual terms on a case-by-case basis.

Research

Though the aquarium has few resources to devote to research on its single dolphin, some research projects are carried out within the facility. One recent vision/echolocation project is studied how dolphins utilize their senses to navigate and identify objects. The Aquarium might consider proposals for similar studies from qualified researchers..

Other Programs

The CMA runs "The Full Circle Program," an animal-assisted therapy program for children with disabilities. (See separate entry on page 338.) The CMA offers an annual family membership for $35, as well as otter and dolphin adoption packages. (A $70 membership gets you a numbered lithograph of a painting by Sunset Sam himself.)

Summer Camp is one of the center's more popular programs. Each week-long class provides students with interactive sessions at the CMA and several trips aboard the center's 26-foot catamaran boat to study local flora and fauna in nearby waters. Past offerings have included "Critter Camp", a fun, creative activity camp for younger kids, and introductory marine biology programs for older children. Other programs are offered during the school-year; contact the education department for current offerings.

The Dolphin Connection

c/o Hawk's Cay Resort
Mile Marker 61
Duck Key, FL 33050
Phone: (305) 289-9975
Fax: (305) 289-0136
Cheryl Messinger - President

Overview

The Dolphin Connection is made up of a collection of natural seawater dolphin pens within the Hawk's Cay resort hotel on Duck Key, Florida. (Duck Key is near Marathon, and is roughly an hour's drive from Miami.) The location has been home to a number of dolphin facilities over the past fifteen years. It was operated until recently by the Chicago Zoological Society as a breeding facility and is now an independent organization with a focus on research and public education.

Though it is not one of the U.S.'s four "swim-with" facilities, TDC offers the public a chance to interact in the water with the facility's three dolphins from several partially submerged platforms of varying depths. (See Other Programs, below.) It is still in its early stages of development as an independent entity, though it already has numerous ties with the scientific and animal care communities. TDC has greatly expanded its educational programs, and is working with the local community to provide a wealth of conservation messages.

Employment, Intern, and Volunteer Opportunities

The Dolphin Connection has only just recently introduced volunteer and intern programs, and considering the organization's potential for growth, its opportunities for involvement may increase in the not-too-distant future.

Internships are open to college students currently enrolled in a biological or zoological field of study, and who are headed for a career in animal care, conservation, or research. Interns are involved in nearly every aspect of the facility's operations, including public interaction, research data collection, cleaning, and food prep. TDC takes an interest in the education of its interns, who are given required reading materials and a project to complete. Interns don't receive much hands-on experience with the animals. A reciprocal agreement with other internship programs sends students to

(Photo by Michael Milford and provided by TDC)

The Dolphin connection is one of a few facilities where people can interact with dolphins in the water.

facilities like the Dolphin Research Center and Theater of the Sea.

Volunteers are welcome at TDC, and are asked to commit to one day per week for six months. Applicants must be at least 18. Volunteers are used in cleaning, maintenance, and food preparation, and generally receive no direct contact with the dolphins. Volunteers are also needed for the group's developing outreach programs. As usual, additional opportunities are afforded to particularly capable and persistent individuals.

Research

There are currently two studies underway at TDC. One is a reproduction study designed to help bottlenose dolphin breeding efforts around the world. The other is still in planning stages, but involves an in-depth look at cranial blood flow in cetaceans, and promises to yield exciting results with practical applications amongst humans.

Other Programs

The Dolphin Connection's "Dolphin Discovery" program allows participants to spend an afternoon learning about the environmental issues facing Florida's marine ecosystems, as well as dolphin anatomy, biology, and husbandry, coupled with a chance to interact with the dolphins from a number of submerged platforms. A less immersive version, "Dolphin Detectives", is tailored to younger visitors, and gives them a chance to see what working around dolphins is all about.

The Dolphin Connection is trying hard to make a difference in its community by offering educational opportunities and swaying its visitors to the cause of conservation. The effort is meeting with great success, and visitor feedback has been almost 100% positive. It will be interesting to see where TDC's programs will take it over the next few years.

The Dolphin Institute & Kewalo Basin Marine Mammal Laboratory

1129 Ala Moana Blvd.
Honolulu, HI 96814
Phone: (808) 591-2121
Fax: (808) 597-8572
http://www.dolphin-institute.org
Dr. Louis Herman - Director
lherman@hawaii.edu
Dr. Adam Pack - Assistant Director
pack@uhunix.uhcc.hawaii.edu

Overview

Kewalo Basin Marine Mammal Laboratory hosts the only dolphin research program aimed at figuring out how smart dolphins are. The Lab's Director, Dr. Louis Herman, has investigated the cognition and communication methods of cetaceans since the early 70's. Over the past two decades the Lab became a sort of Mecca for students interested in dolphin intelligence. While four Atlantic bottlenose dolphins live in the Lab's dolphin habitat, it is not open to the public, and would otherwise have been listed as a research facility. The Lab offers students and interns practical experience in both dolphin husbandry and training, and cetacean research.

KBMML is in the process of relocating its facilities to make way for a public park expansion. "The Dolphin Institute" is a not-for-profit organization designed to locate a new home for the Lab's animals and staff. Once the program relocates, the whole thing will be called The Dolphin Institute. Keep your eye on the Omega website for updates.

Employment, Intern, and Volunteer Opportunities

There are very few paid positions at the Lab, but there are many opportunities for interns, and the it takes on a limited number of graduate and undergraduate students through the University of Hawaii. KBMML also cooperates with Earthwatch, giving the general public a way of participating in its programs. Take a look at the Earthwatch program on page 170. If you're interested, contact KBMML directly.

Competition for internships is even more intense here than at other marine mammal facilities. Internships last an entire semester or summer, and are a full-time commitment. Interns usually become involved with one or more dolphin research projects, and learn research techniques as well as necessary animal care and training skills. Fish preparation, habitat maintenance, and cleaning are just as much a part of an internship as data collection, recording, and analysis. Applicants must have at least two years' worth of college experience, and course credit can be arranged either through the student's host school or through the University of Hawaii. Internships begin in January, May, and August, and application deadlines are October 1st, February 1st, and May 1st, respectively. Applications should include all college transcripts, three letters of recommendation, preferably from professors with a personal knowledge of the applicant, and a personal statement outlining the applicant's background, skills, objectives, and expectations.

Opportunities for graduate students are offered through the University of Hawaii's Department of Psychology, under the Human and Animal Cognition Program. The program supports a maximum number of students of roughly twelve. While designed primarily for psychology students, students majoring in biology, zoology, and computer science have been accepted as well. Acceptance is based on transcripts, GRE scores, letters of recommendation, and evidence of prior schoolwork and experience relevant to the Lab's programs. There is particularly heavy competition for KBMML graduate positions.

Almost as rare but providing an excellent way to get your foot in the door are undergraduate courses taught by Dr. Herman himself at the University. Participants assist with research and operations at the Laboratory at least three half-days each week, and are essentially involved in the same work as interns. The course may be repeated as well, giving undergraduate students considerable experience in research, husbandry, and training techniques.

Research

The Lab's researchers have conducted dozens of studies with the facility's dolphins, many of which have focused on the use of symbolic languages to give instructions, and to a lesser degree to even pose questions to the animals. While these efforts have not produced a magic intelligence meter stick by which we might finally judge cetacean intelligence and compare it to our own, the group's experiments show us a great deal about the cognitive abilities, and limitations, of dolphins. For more information about the results of KBMML's research, see the discussion of cetacean intelligence in chapter 3.

The Lab also conducts social behavior research on the North Pacific humpback whale. KBMML pioneered humpback whale research in Hawaii in 1975. The study continues today, tracking humpbacks all the way from their summer feeding waters in Alaska to their mating and calving waters near the Hawaiian islands.

✱ Dolphin Quest

68-425 Waikoloa Beach Dr.
Waikoloa, HI 96738
Phone: (808) 886-1234x1505
Fax: (808) 885-7030

The one feeling I couldn't shake during my visit to the Hilton Waikoloa on the big island of Hawaii was that at any moment Ricardo Montalban was going step out from behind one of the massive pillars in a white suit and offer to turn me into a movie star. Sadly, this did not happen, but the resort had much else to offer. The place is ridiculously luxurious, bordering on extravagant. I loved it. It's actually more like a collection of hotels, since it's made up of several buildings' worth of suites, shops, restaurants, and musical combos. There are many ways to get around the compound, including a small modern-looking electrical train, a flotilla of boats which make regular stops along the little river which winds through the place, and a series of paved paths which meander amongst copses of hammock-strewn palm trees and oddly shaped, multi-level swimming pools. At night the place is almost magical (gag - I thought I'd never use that word) with the soft rustling of surf and warm, gentle breezes teasing the flames of torchlights everywhere. If you think this sounds like a pleasant place to work, you're probably right! Of all the facilities in Hawaii, Dolphin Quest seemed by far the most appealing.

Overview

Nested in the center of the Hilton's affluent sanctuary is the natural lagoon housing the resident Atlantic bottlenose dolphins. The lagoon is fed directly by ocean water which passes through the island's porous lava rock, creating a natural filtration system as good as any man-made sand filter. Dolphin Quest is an international company and maintains two other facilities in Tahiti and Bermuda. The Waikoloa facility, which was built in the mid 80's by two marine mammal veterinarians, is treated as a contractor by the resort, which leaves the staff free to design their programs around the animal's needs while still providing exceptional service to the hotel's guests.

Human-dolphin encounters are really what Dolphin Quest is all about, and they provide four main types of programs. "Dolphin Discovery" is aimed at children ages 5 to 12, and is a very safe shallow-water session in which a small group of kids, accompanied by a trainer, are brought into contact with a dolphin under controlled circumstances.

A similar program exists for teenagers, and a more interactive program accommodates adult guests (16 and older) who have an opportunity to enter the deeper part of the lagoon for a controlled "swim-with" session. A fourth program caters to couples, and takes place in the shallow area. All programs include excellent educational content commensurate with the age group, and guests have the option of purchasing a videotape of their experiences which is edited together on-premises in Dolphin Quest's editing studio.

Trainers are present at all times during the encounters, and the animals themselves always have the option of retreating to a "dolphin only" area, which makes for a very safe and responsibly run program. Prices range from $45 to $130, depending on which you choose. Child and teen programs may be reserved up to two months in advance, but due to the high demand, adult encounters are selected on a daily basis at the resort by lottery.

Employment, Intern, and Volunteer Opportunities

Dolphin Quest does enlist the aid of volunteers, though they act primarily as docents, and consist of local residents who are looking for an interesting pastime, as opposed to a potential career path.

Internships are available for college students, and positions open up roughly once a year or so in DQ's full-time staff. Most interns at Dolphin Quest are students from Moorpark College (see separate entry) who are fulfilling their 2 week internship requirement for graduation. Interns from other schools are welcome to apply, but it has been difficult for past applicants from the mainland to make arrangements for a potential internship. Interviews are conducted in person at the applicant's expense, and should they be accepted they must then contend with living arrangements. Hawaii is, after all, one of the most expensive places to live in the Milky Way Galaxy, and the internships are unpaid (except for a $50/week stipend - not enough to rent a shoebox in Waikoloa.) Students from the University of Hawaii stand a much better chance of snagging an internship, particularly if any of Dolphin Quest's animals are in the latter stages of pregnancy, which usually places a higher demand on the staff.

Internships are available for students from Moorpark College who are fulfilling their two-week requirement for graduation.

Internships are also offered during times when dolphin births are impending. Interviews for paid positions are onsite and last between one and two weeks. (So bring a bottle of water along with that resume.) During this time, Dolphin Quest supplies the applicant with a place to stay, as well as transportation to and from the airport. Applicants must provide their own food and airline tickets. This extended interview process gives potential employees a chance to see what Dolphin Quest is all about, and to interact with the staff. Applicants who successfully complete the interview process may be brought on staff for a three-month probationary period.

There are over 20 permanent staff members at Dolphin Quest Hawaii, most of whom are cross-trained to work in all areas of operation. This includes training, productions, retail, and office work. Entry-level trainees start out primarily in video production as a convenient segue into animal care. By taping the sessions (which requires them to enter the water with the groups) they grow accustomed to being in the water with the animals, and get a good feel for procedure. Dolphin Quest has hired a number of Moorpark graduates, and prefers to hire entry-level applicants over experienced trainers.

One way around the initial cost of traveling to Waikoloa for an interview is to track down Dolphin Quest staff at the annual IMATA conference (see separate entry.) If the timing is right, you might even be able to arrange in advance to meet with a staff member at the gathering, which usually takes place in the second week of November.

Dolphin Quest staff work 5 days a week, typically from 8:15 to 5:15, with excellent benefits. A nice perk is the opportunity for staff members to work for extended periods of time at their sister facilities in Tahiti and Bermuda. (Life's rough, eh?) As if that weren't enough, lunch is provided, gratis, by the hotel, and is generally regarded as fantastic.

Other Programs

Dolphin quest provides numerous informational handouts to its visitors on a wide variety of marine issues, including conservation and natural history. It also donates a sizable portion of its income to the Waikoloa Marine Life Fund to be used for researchers who apply for funding. Inquiries about the Fund can be sent to Dolphin Quest's address.

Dolphin Research Center

P.O. Box 522875
Marathon Shores, FL 33052
Phone: (305) 289-1121
Fax: (305) 743-7627
http://www.florida-keys.fl.us/dolphinresearch

The Dolphin Research Center has borne four different names over the years, and began its life as "Santini's Porpoise School" in 1958. Legend has it that Milton Santini, a local fisherman, inadvertently snagged a bottlenose dolphin in one of his nets, and placed it in a large tank to nurse it back to health. He became fascinated with the animal, and before long had gathered a collection of dolphins for public display. One of these dolphins, Mitzi, was the first of many to play the role of "Flipper", which she did in the pilot movie of the same name. The facility changed hands in 1971 to become the dolphin show "Flipper's Sea School", and again in 1977 as the "Institute for Delphinid Research." It was under this name that whale conservationist Jean Paul Fortom-Gouin conducted studies on the language and reasoning abilities of dolphins in an attempt to bring about an end to commercial whaling. With the passing of the international whaling moratorium in 1983, the facility was handed over to its senior trainers, and the Dolphin Research Center was born.

Overview

DRC is one of the half-dozen or so cetacean groups which dot the arc of the Florida Keys, and is also the most popular amongst those seeking a career with marine mammals. The Center is actually a small compound on Grassy Key, roughly halfway between Key West and the mainland. The Center was established in 1984 for the purpose of "teaching, learning, and caring for marine mammals and the environment we share." Active in its local community as well as in national and international research, therapy, and rehabilitation circles, DRC has become synonymous with dolphin care and interaction throughout much of southern Florida. It is one of four facilities in the U.S. licensed to run "swim-with" programs, and has become something of a minor Mecca for students and children looking to learn about, and interact with, dolphins.

Employment, Intern, and Volunteer Opportunities

Offering practical experience to volunteers and interns is something at which DRC excels. They possess one of the most comprehensive internship programs in the country, and offer a wide variety of opportunities to volunteers.

Internships at DRC are designed for college students, though others may apply. Interns are accepted three times a year, in the summer, fall, and winter. Internships are unpaid, last from three to four months, and are offered in six of the Center's departments. Animal Care and Training interns assist with dolphin and sea lion sessions, interact with the public, assist with research data collection, and maintain equipment. Animal Husbandry interns assist with food and vitamin preparation, stocking food and husbandry supplies, fish deliveries, record keeping, research data collection, and medical procedures and observations. Interns with the therapy group maintain equipment, assist program participants and their families, record data, and handle program inquiries and correspondence. Education Internships involve administrative assistance with the Dolphinlab program (see below), library maintenance, and research project data collection. Medical care interns help the medical staff with their duties, and research interns are intimately involved in the execution of research projects.

DRC makes particularly effective use of staff/intern relationships. Interns are paired up from their first day with a staff member, who serves as a coach and supervisor. The intern and coach collaborate on a list of educational goals, and meet regularly to discuss the intern's accomplishments. One tool used to chart the intern's progress is the daily journal they are required to keep.

Shorter versions of these internships, called "Career Development" internships, are available on a limited basis for students unable to commit for more than four to six weeks. All interns assist with general volunteer duties, including food preparation, cleaning, and various administrative tasks. There is usually only one position open per season in each department.

DRC makes use of local volunteers as well as those from out of town. Non-locals are asked to commit to a four to eight week stay, and generally work 30 to 40 hours per week. These volunteer positions are not as structured as DRC's internships, and are open to anyone with an interest. Local volunteers are welcome to help out once a week.

Volunteers often face many different tasks, and are made use of by almost every department. Food preparation, cleaning, the feeding and care of exotic birds, trash collection, painting, and landscaping are all likely activities for a DRC volunteer. Few requirements are made of applicants, though they must be able to speak and write in English. Children under 18 may apply, but must be accompanied by an adult.

Both volunteers and interns can apply for access to DRC's limited number of dormitory-style housing, adjacent to the main compound. A bed goes for $390/month, or $130/week, and requires one month's payment and a $100 security deposit in advance. Bed linens, bath and kitchen access, local telephone service, and use of a coin-operated washer/dryer is included. Housing is not guaranteed, and preference is given to full-time participants.

The Dolphin Research Center has a staff of roughly 40, about a dozen of which work with the marine mammals. Positions open up once every year or so, and DRC has filled a number of vacancies with prior interns and volunteers. The animal care staff has more responsibilities at DRC than at many display facilities. In addition to caring for the animals, the staff is involved with the Dolphinlab, Dolphin Insight, and Dolphin Encounter programs (see below), as well as being called upon to guide interns, rescue and rehabilitate stranded manatees, and coordinate research projects, all of which makes for a very full plate in a business where free time and energy are already precious.

Research

Research projects are constantly changing at DRC, though it has several with long-term goals. An ongoing behavioral study of dolphin calves has been keeping interns and volunteers busy for years, requiring intensive daily observations of a newborn dolphin for every day of its first year. Several ongoing studies are investigating dolphins' cognitive abilities, including cooperative behavior tasks, object discrimination, and match-to-sample studies, where a dolphin is asked to pick out an object from a series. Volunteers and most interns have opportunities to assist in these and other research projects.

Other Programs

While DRC used to be very active in cetacean strandings, it is now focusing its rescue and rehabilitative efforts exclusively on manatees, due mostly to the presence in the Keys of other volunteer groups able to respond to cetacean incidents. DRC is the only facility in the Keys permitted to assist the endangered manatee. In addition, the recent appearance of the highly contagious morbillivirus in wild cetaceans puts DRC's colony at risk when staff work with potentially infected animals. (Manatees carry no such risk.)

For those willing to kick out for the $1,000+ price tag, DRC's "Dolphinlab" program offers seven days of intensive exposure to marine mammal education, including a fair amount of hands-on,

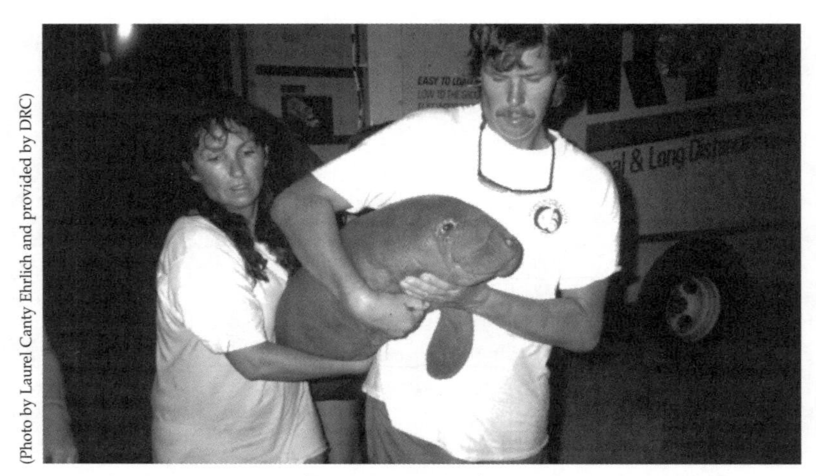

(Photo by Laurel Canty Ehrlich and provided by DRC)

A manatee rescue in progress. Ryder trucks make manatee transportation safe, efficient, and smooth. Ryder - the right choice.

interactive experience with the Center's dolphins. A battery of seminars explores cetacean physiology, marine mammal legislation, stranding response, training, dolphin-assisted therapy, bioacoustics, and behavioral, cognitive, and biological research. The tuition covers dormitory lodgings and food for the week, and transportation to and from Marathon airport. Dolphinlab can be taken for college credit, either through the student's college or the nearby Florida Keys Community College.

Intermediate and advanced versions of Dolphinlab are available, which are open to graduates of the Basic course. The Intermediate level is a more in-depth look into marine mammal conservation and research, while the advanced course is designed to give participants a chance to focus on one particular topic. In terms of learning animal care and training techniques, this could be very useful for young adults who are hell-bent on breaking into the field. Applications should be submitted at least three months in advance.

"Dolphin Encounter" is DRC's swim-with-the-dolphins program. The program is offered to adults, and children ages 5-12 when accompanied by an adult. Reservations are required and are taken only by phone. Bookings for a given month begin on the first of the preceding month. The program is immensely popular, and an entire month will often fill up in one day. Cost for the program is $90 per person; family members may observe for $15 each.

DRC also offers "Dolphin Insight", a less involved version of "Dolphin Encounter" where participants interact with the dolphins from a platform, instead of being immersed in the water. It also runs the "Dolphin/Child Program", which is dolphin-assisted therapy for disabled or troubled kids. (See the description on page

(Photo by Laurel Canty Ehrlich and provided by DRC)

Dolphinlab gives participants a good introduction to dolphin biology, animal care, and natural history.

Dolphins Plus

PO Box 2728
Key Largo, FL 33037
Phone: (305) 451-1993
Fax: (305) 451-3710
http://www.pennekamp.com/dolphins-plus
dolphins-plus@pennekamp.com

Overview

Dolphins Plus is one of four cetacean facilities in the United States which are permitted to run "swim-with" programs, and was the first such facility to do so. The facility consists of a number of natural lagoons which are fed directly from the ocean. Dolphins Plus houses about twenty bottlenose dolphins which are made available to the public through educational swim sessions, dolphin-assisted therapy sessions, and classes.

Dolphins Plus has an excellent track record for animal health and successful rehabilitation of stranded animals. (Indeed, they are the only display facility to have never lost an adult dolphin.) It is worth noting, however, that the facility's policies and philosophies towards animal care are unique, and work experience at Dolphins Plus may not transfer easily to other cetacean display facilities.

Employment, Intern, and Volunteer Opportunities

There are no volunteers at Dolphins Plus. The facility supports a stranding response group, the Marine Mammal Rescue Service of the Upper Keys, which does enlist the aid of local volunteers on an on-call basis.

Internships are offered, but must qualify for college credit at the student's home institution. The student and a sponsoring professor must design a research project with Dolphins Plus' approval. Past projects have focused on areas like behavioral research of mothers and calves, and generally will not bring the intern into direct contact with the animals.

There are about 10 paid animal care workers at Dolphins Plus. No premium is placed on prior marine mammal experience in potential applicants. Positions open up very rarely, however, and in the past, years have gone by without any new trainers being hired.

Research

In addition to the projects conducted by previous interns,

Dolphins Plus continues to add to the growing base of information concerning dolphin-assisted therapy. Dolphins Plus helped pioneer this type of program in the United States, and provides a valuable service to therapists of behavior-disordered and autistic children worldwide. Dolphins Plus has been kind enough to offer its services free of charge to selected local children who would otherwise be unavailable to participate. They are currently hosting the Island Dolphin Care program (see page 339.)

Other Programs

Sessions at Dolphins Plus cost about $85. This incudes an hour-long educational and safety briefing followed by a half-hour swimming session with the dolphins. The dolphins used in these sessions aren't trained, so there is no guarantee of contact with the animals: it's entirely up to them. Children must be over ten years old, and anyone under eighteen must be accompanied in the water by an adult. Participants must be adept swimmers since the session takes place in deep water, and flotation devices are not allowed. A "half day" program is available for persons wishing to double their time in the water. For children under ten, or for persons wishing guaranteed contact with the animals, a controlled swim with trained dolphins is offered on a more limited basis.

Three and five-day marine mammology courses are offered at Dolphins Plus which provide students with a basic grounding in cetacean evolution, anatomy, environmental issues, care of captive marine mammals, physiology, dolphin human therapy, and coral reef ecology. The cost of the program varies based on the number of attendees, and schedules are preset about 2-3 months beforehand. Contact the education department for details.

Epcot Living Seas

P.O Box 10,000
Lake Buena Vista, FL 32830
Phone: (407) 560-6922
Fax: (407) 827-8790

Think huge. I mean really huge - huge with all the trimmings. Remember that thing you saw you thought was big? It's bigger than that. Conservative alarmists are probably worried that Walt Disney World is going to declare itself an independent nation, it's so big. Most people will tell you Disney World is "in" Orlando. At just under fifty square-miles in area, the park is about as big as Orlando! The biggest, baddest theme park on the planet also has the world's biggest indoor dolphin habitat. It isn't easy to work at EPCOT Center's Living Seas complex, but if you're after the primo job in cetacean care and training, this might be for you.

Overview

Disney is first and foremost a professional organization, obsessively concerned with its image and the quality of its product, whether it be feature-length films, consumer merchandise, or the experience of visiting one of its theme parks. Being a "Cast Member" at one of the Disney parks is a demanding job, no matter what function you serve. Employees interact with guests according to strict guidelines, and are expected to be neat, punctual, and courteous. There are at least three separate publications which outline in great detail acceptable hairstyles, hair colors, clothing styles, fabrics, jewelry, make-up, shoes, hats, sunglasses, and hosiery. Individuals with visible tattoos, beards and mustaches, or completely bald heads do not meet set standards for the "Disney Look".

The result of such specific and comprehensive policies is unparalleled consistency and an impeccable professional image throughout Disney's theme parks. Though Disney has an unmistakably corporate feel to it from the inside, it knows the entertainment business better than anyone, and the quality of its products remains unmatched.

This is certainly true of EPCOT Center, one of the three main theme parks comprising Walt Disney World in Orlando, Florida. The park is divided into the Future World and World Showcase sections. World Showcase is a collection of eleven miniature "cities" representing food, shops, and entertainment from foreign countries, including Norway, Morocco, France, Italy, and China. The mini-cities are staffed by foreign students, lending an air of authenticity to the attractions. In Future World Disney features a host of technological exhibitions themed around the human body, automobiles, communication, energy, and among other things, the oceans.

The Living Seas exhibit takes visitors inside "Sea Base Alpha", a research base supposed to be resting on the ocean floor. The designers and architects really went to town trying to make the place look futuristic - I kept expecting Geordi LaForge to pop in and run a bypass or something. The enormous building is festooned with hatches, airlocks, glass tubes, and funky decals on the walls. A multitude of educational exhibits touch on a wide variety of marinelife topics. The main tank is filled with nearly six million gallons of seawater, and is home to a gigantic simulated coral reef, including dozens of species of fish, rays, and sharks, as well as half a dozen bottlenose dolphins. Interestingly enough, Disney downplays the dolphin's presence, omitting them altogether from the park's literature and website, causing most guests to be pleasantly surprised.

There are no public presentations with the animals, though staff members regularly narrate training and research sessions. The tank is amazingly huge, and is circular in nature, permitting the staff to section off various parts of it with movable nets radiating from the center. Though impressive when seen from above, the habitat can only be seen by the public through large glass windows, adding to the "seabase" illusion. In separate habitats, manatees can be viewed from above, where staff members conduct Q & A sessions throughout the day.

Employment, Intern, and Volunteer Opportunities

There aren't any volunteers at Disney, but they do have a couple of very good internship programs. Technically, Living Seas interns are part of the Disney College Program, which recruits college students for a multitude of service and administrative tasks. Consequently, potential applicants must complete an interview at a local college or university with a Disney representative. (Reps travel all around the country to more than one hundred schools each year.) Interviews are conducted in the fall for January positions, and in the spring for positions beginning in May and September. Times and locations of interviews are available by calling (800) 722-2930. In this regard, standard interviewing skills are more important in pursuit of an internship at Disney than at most other facilities, since your first cut must be made before the animal staff even sees your name. Disney looks for people who maintain eye contact and possess excellent communication skills.

There are two kinds of marine mammal internships at The Living Seas; one is unpaid and lasts just under four months, while the other is a six month long paid position. The six month internship focuses more on animal care and training. Much time is spent cleaning, preparing food, interacting with the public, maintaining habitats with SCUBA gear, and record keeping. These interns participate in research sessions and feedings for the dolphins and the manatees. Participants have the opportunity to qualify for specialty SCUBA certifications. There are generally four of these interns taken on each year; two begin in late December and two in mid-June. Interviews and applications should be completed six to twelve months prior to the desired starting date. The program is aimed specifically at college juniors, though some flexibility may exist for interested seniors and sophomores. First aid, CPR, and open water SCUBA certifications are required of all candidates. Applicants should be able to demonstrate some kind of experience in marine science or education, whether through coursework, extracurricular activities, or volunteer experience at another facility. The internship pays just over $300 per week, and a modest relocation allowance is provided. Housing at Disney's Vista Way apartment complex (which by the way is VERY nice) is a bargain at roughly $75/week. Interns are housed two to a bedroom in two and three bedroom apartments, and transportation is provided to and from work.

The unpaid internship focuses on the dolphin research program (see Research, below.) Interns assist the research staff with data collection and entry, maintain and operate the project's equipment, set up equipment underwater with SCUBA gear, and also narrate research sessions for the public. There are about twelve of these internships offered each year in groups of four, each group corresponding to an academic semester (spring, summer, and fall.) Interviews and applications should be completed at least two to three months prior to the desired semester. These interns work closely with the primary researchers, and attend research planning meetings, giving them excellent exposure to formal research in an enclosed environment. The position is open to all undergraduates, and recently graduated students pursuing a graduate degree. Applicants are expected to have some experience or coursework in experimental or animal psychology, and to have a basic understanding of operant conditioning. A SCUBA certification is helpful, but not necessary. Although this internship is unpaid, participants do have access to the Vista Way apartment complex as a low-cost alternative to finding housing on their own.

There are eight full-time staff members in the marine mammal group. Each trainer works with both the dolphins and the manatees. Though there are no public shows with the animals, the staff spends a lot of time with the visitors, explaining what's happening during training and research sessions, and answering questions. They spend a lot of time in the water, more so than at most other facilities, and have the most extensive collection of SCUBA equipment I've ever seen. Permanent positions open up infrequently at The Living Seas, about once every three or four years. When the positions open up, (and you pretty much have to have an inside contact to know when they do) they are generally filled by prior interns or experienced trainers from other facilities.

There is also a full-time dive staff at The Living Seas. These staff members spend most of each day maintaining the habitats in SCUBA gear, showing off the facility's expensive submersibles and helmeted dive gear, and doing The Tube Thing. The Tube Thing is where, once every half hour, one of these guys enters the top of a gigantic water-filled tube in the center of the public area and descends into the room to the oohs and ahs of the gaping public. The tube is then drained and the diver crawls out of a small hatch to talk to the visitors about the tube, his dive gear, etc. It looks really neat the first time you see it, but I imagine the staff must get tubed-out pretty quickly. Anyway, there are about eight of them, with positions opening up every couple of years.

Working at EPCOT can be a bit of a surreal experience, even weird. Once you get used to seeing giant ducks and mice walking around every day, there's still an otherworldly feel to the place, which I suppose is the point. Go and visit the park as a guest, and see if you can arrange to speak to a staff member beforehand. If you can get a feel for what the place is like, and what is expected of the "cast", you can better judge whether it's the right place for you.

Research

The Living Seas spends more time conducting cetacean research then almost any other display facility in the nation. Previous studies, conducted by outside researchers, have focused on dolphin cognition, echolocative ability, acoustic perception, tool use, and problem solving. Right now, it is one of only two facilities in the U.S. actively attempting to determine whether cetaceans have the potential to learn and utilize a complex, syntax-rich language similar (yet simpler) to our own. Led by investigators John Gorey and Mark Xitco, researchers have constructed an enormous "keyboard" for the dolphins to use in conversation with humans. This sort of thing has been done before with primates, and is somewhat similar to the work done at The Dolphin Institute (entry on page 61.) For more information on this program, see the cetacean intelligence section on page 19.

Other Programs

Visiting school groups may arrange to have a staff member speak about the facility and its residents. Also, certified SCUBA divers can arrange to go diving in the main habitat - separate from the dolphins - for about $140. The session is a three hour process and includes a 30 minute dive.

It's always a good thing when a colossal corporation like Disney adopts sound environmental policies. Any organization receiving tens of millions of visitors each year faces a mountain of environmental issues. Luckily, Disney recycles like mad, having built an enormous Material Recovery Facility in 1992. Each day 14 TONS of reusable plastic, paper, glass, steel, aluminum, and cardboard are extracted from Disney's garbage, and 25 tons of compost are produced from sewage, landscape waste, and wooden pallets. Used equipment, leftover building supplies, computer components, and old costumes are sold off or donated to local charities. Each year Disney donates around 240 tons of prepared but unused food to charity, and recycles 3,000 tons of food waste as livestock feed and compost. The company uses recycled materials for its brochures and shipping containers, and has worked hard both to reduce pesticide use and to replace harmful pesticides with environmentally safer chemicals. Disney's Environmental Initiatives department invites queries, and can be reached at the main address above.

Gulf World

15412 Front Beach Road
Panama City Beach, FL
32413
Phone: (904) 234-5271
Fax: (904) 235-8957

Overview

Gulf World offers family-oriented entertainment in the form of dolphin shows, Sea Lion acts, petting pools, performing parrots, and underwater demonstrations. It houses bottlenose dolphins, sea lions, penguins, parrots, river otters, turtles, sharks, and American alligators in a series of outdoor habitats. Gulf World will be more than doubling its size in 1998 with the addition of new dolphin and parrot stadiums, an enclosed tropical garden, and a 50,000 square-foot shopping attraction entrance.

Employment, Intern, and Volunteer Opportunities

Gulf World does not use volunteers, and has no official internship program. Plans for implementing internships are underway, and may be in place by the end of 1998. The park is open from February through October, and during this season it does hire temporary employees. Seasonals are often local high school or college students, and Gulf World has often hired applicants with little or no experience. New employees usually begin in a support role and work their way up to animal care over the course of two or more seasons.

There are about half a dozen permanent staff members in the marine mammal department, all of whom had prior experience either as a Gulf World Seasonal employee, or as an animal care worker at another facility. When the additions to Gulf World's facilities are completed, the park will be hiring a large number of animal care workers, so keep your eyes open toward the end of 1998.

Other Programs

Gulf World is an active member of the Marine Mammal Stranding Network, and has the facilities to rehabilitate small cetaceans. The park doesn't actively recruit local volunteers to help respond to strandings, but could probably use help when caring for a stranded animal.

Gulfarium

1010 Miracle Strip Parkway
Ft. Walton Beach, FL 32548
Phone: (850) 243-9046
Fax: (850) 664-7858

Overview

Founded in 1955, the Gulfarium is one of the country's oldest marine parks. It sits in a touristy stretch of Okaloosa Island, just a stone's throw from downtown Ft. Walton Beach. It consists of an outdoor compound made up of various marine, avian, and terrestrial animal exhibits, which are presented to the public in a variety of formats with an emphasis on entertainment. The Gulfarium is home to bottlenose dolphins, sea lions, penguins, seabirds, river otters, turtles, and American alligators.

Employment, Intern, and Volunteer Opportunities

The Gulfarium does not use volunteers, though it does hire seasonal interns in the summer. The Gulfarium has a history of hiring young and inexperienced applicants to fill its seasonal positions, which can often turn into a permanent job in animal care. New hires usually begin as divers, helping with habitat cleaning and maintenance as well as the park's underwater animal sessions. Previous interns have come primarily from the University of Florida and Santa Fe Community College, though others may apply. The Gulfarium has also taken on local high school students as young as fifteen years of age, under close supervision.

Other Programs

The Gulfarium is an active member of the Marine Mammal Stranding Network, and occasionally provides rehabilitation facilities for cetaceans. The Gulfarium is still caring for a spinner dolphin, named Kiwi. Having been brought back to health but deemed unreleasable, Kiwi has become the park's newest attraction. "The Kiwi Experience" is an opportunity for the public to sit in the water with the dolphin, two at a time, under the supervision of a trainer. The Gulfarium does not currently accept stranded animals for rehabilitation until the animal has been given a clean bill of health by the park's veterinary staff.

The Hawaii Institute of Marine Biology

P.O. Box 1346
Kaneohe, Hawaii 96744
Phone: (808) 236-7401
Fax: (808) 236-7443
http://www.soest.hawaii.edu/HIMB/himb.html

Overview

The Hawaii Institute of Marine Biology, is a subdivision of the School of Ocean and Earth Science and Technology at the University of Hawaii. The Institute's facilities are both considerable and beautiful. Twenty-six kilometers from the main campus of the university, Coconut Island sits in Oahu's Kaneohe Bay, tucked behind an extensive barrier reef system which contains a wide variety of marine life. Outside the barrier reef, pelagic species and open-ocean waters up to a thousand meters deep are less than six kilometers away.

When the Navy pulled its marine mammal program out of its Hawaii facility several years ago, a number of marine mammals and researchers remained at HIMB at the University of Hawaii's request. The University has since formalized its Marine Mammal Research Program, and the Institute now houses an assortment of bottlenose dolphins, Risso's dolphins and false killer whales. The animals are not accessible to the public, and the Institute would otherwise appear in the Research section of this book.

The facilities at HIMB include fish holding tanks, underwater video recording and data analysis gear, a darkroom, University mainframe computer terminals, a SCUBA equipment room and air compressor, a maintenance and engineering shop, and six controlled tidal ponds with a total area of nearly five square kilometers. Also at the lab's disposal are the twenty-seven foot research vessel Mao Mao, a thirty-five foot passenger and cargo number named Hono Kai, and an assortment of skiffs.

The Institute has office and laboratory facilities in the Marine Science Building on the University campus, and operates the Mariculture Research and Training Center at the northern end of Kaneohe Bay. Like the name says, this site is used for aquaculture research and includes laboratory, shop, office and dormitory buildings, a dozen earthen ponds totaling eight acres, dual airbags, salt and freshwater capabilities and a fully automated data acquisition system.

Research Projects

Most of the Marine Mammal Research Program's efforts center on either echolocation or hearing capabilities of cetaceans. HIMB also conducts research in the areas of marine animal behavior, coastal environmental studies, coral reef biology and ecology, fisheries biology, fish endocrinology, and tropical aquaculture.

Involvement Opportunities

The HIMB Marine Mammal Research Program can support roughly four graduate students at a time, under the guidance of University of Hawaii professors Dr. Whitlow Au and Dr. Paul Nachtigall (the Program's Director). The Program also makes use of a number of undergraduate volunteers and is actively involved with minority student education with an emphasis on Pacific Island people.

Another group worth looking into is Marine Education and Research on Coconut Island, a part of the University of Hawaii Foundation. MERCI's mission is "to enable pursuit of new research initiatives and to provide scholarships and other support for scientific endeavors by students, researchers, and educators at HIMB." I honestly don't know what that means in practical terms, but in theory, this is the sort of group to which ambitious students and researchers submit proposals for funding. MERCI circulates a newsletter amongst its members, and runs a number of activities on the island, including a docent program, lectures, tours, workshops, and fundraising. Correspondence for MERCI should be sent to the Institute's address.

Indianapolis Zoo

1200 West Washington Street
Indianapolis, IN 46222
Phone: (317) 630-2140
Fax: (317) 630-5153

Overview

The Indianapolis Zoo, possibly the largest Indy attraction having nothing to do with driving cars around and around in circles, is parked just west of the downtown area of the city. The original Indianapolis Zoo was the brainchild of Lowell Nussbaum, a local newspaper columnist in the '30s and '40s. Although he founded the Indianapolis Zoological Society in 1944, it took two more decades before the Zoo actually opened. Originally designed as a children's zoo, the facility was built entirely on revenue from corporate, foundation and individual donations. To this very day, the Indianapolis Zoo is completely self-sustaining and is one of the few zoos in the United States to operate completely without tax support.

The Zoo was popular and well attended from the start, and it took less than 20 years for it to outgrow its boundaries on the east side of Indianapolis. With a successful capital campaign, a new and expanded Indianapolis Zoo was opened in 1988 in the 250-acre White River State Park. Joining the trend of naturalistic zoo habitats, the Society created biomes that simulated landscapes of many different natural environments. Not only does this approach offer quality exhibition space for the animals, it helps educate visitors about the importance of habitat preservation. The 64-acre Zoo is always bustling with activity, playing host to social gatherings, education classes, seminars and special events throughout the year.

Volunteer, Intern and Employment Opportunities

The Indianapolis Zoo has one of the most flexible marine mammal departments from a career-seeker's standpoint. Their volunteer and intern programs are designed to be symbiotic relationships benefiting both Zoo and individual. Their full-time positions require only a high school diploma, the ability to become SCUBA certified, and a willingness to interact with the public. As always, college degrees, prior animal care experience, and SCUBA certifications would be helpful. There are just over a dozen full timers in the marine mammal department, all of whom take turns working with the Atlantic bottlenose dolphins, harbor seals, sea lions, walrus, and polar bears (yes, polar bears!) in their care. Staff members are assigned an animal area in which to focus but will usually assist in all areas throughout the week. They also rotate focus areas on a roughly annual basis, so there is quite a bit of cross-training. Public speaking is not required of staff members until their second year or so, but most wind up learning sooner. Quite a few Indy staff members have been hired without prior experience. Many of them were ex-volunteers or interns at the Zoo.

Each school semester, the marine mammal department brings on two interns on a volunteer basis. Unlike many internships, concurrent enrollment in a college is not required, which can be very helpful for recent graduates. The internship lasts for a bit more than two months and is full-time. Of particular interest is the full scope of an intern's responsibilities. If they measure up after the first week or two, they will assume roughly the same tasks as an entry level keeper, working directly with the animals. This alone makes an Indianapolis Zoo internship an invaluable experience.

There are four types of volunteers who might assist the marine mammal department: regular volunteers, bionaturalist volunteers, "ZooTeens", and dive volunteers. The most likely to wheedle their way into a job are the regular volunteers. These people show up one day a week for about four hours. The department's flexibility shows through again in its volunteer program. The staff prefers help in the morning and on the same day each week, but will do its best to accommodate people with conflicting or fluctuating schedules. Regular volunteers assist with food preparation, cleaning, and occasionally assist trainers in sessions. Experience and merit will garner greater responsibility and opportunities for this type of volunteer.

The bionaturalist volunteers' primary function is to interact with the public as docents, although they occasionally assist staff members as well. The dive volunteers are few in number, and the future of this program is uncertain. These volunteers assist primarily in cleaning the underwater habitats. Obviously, a SCUBA certification is needed.

Finally, the ZooTeen program is an attempt to give local students a chance to get a close-up feel for zoo careers. Eligible teens apply by February, with training beginning for selected students on weekends in April and May. Once the program begins, the ZooTeens work one eight-hour day per week all summer long. The program accommodates up to 75 students per year, about of quar-

(Photo provided by the Indianapolis Zoo)

Three of the zoo's Atlantic bottlenose dolphins.

ter of whom end up in the zoo's Dolphin Pavilion. Every year, the list of students wanting to participate gets longer, and the competition more fierce to acquire these interesting positions.

Research

While research plays an important role in the life of the Indianapolis Zoo, much of the cetacean research involves remote analysis of blood and tissue samples that is not readily accessible to research assistants. However, the Zoo opened a new veterinary hospital complex in 1997 that is wired for distance learning (to Purdue University's veterinary school, for example), along with other key areas in the Zoo. The role of research will undoubtedly increase at the Zoo, both through the veterinary staff and the curatorial staff.

Other Programs

This Zoo's education programs were some of the hardest for me to sort through, since there are so many of them! Children, teens and adults have access to scores of experience with zoo careers, natural history and conservation, both at the Zoo and through outreach programs. Teachers receive special training and instructional materials, and they may also borrow hands-on animal materials, informational packets and videotapes. Classrooms may participate in the Zoo's ground-breaking work in distance learning, where a two-way video link brings the class right into the animals' homes. Students are able to join the Zoo staff in otherwise off-limits activities such as checking for baby wallabies in the mothers' pouches! One-hour student workshops are offered on a wide and expanding range of topics, from the Amazon to the zebra. Special behind-the-scenes workshops also are available on a more limited basis. Overnight functions are available in both school group and teacher-only flavors. "Shadow keeper" opportunities let high school students get a day-long taste for zookeeper's life through observation and verbal exchange with a staff member.

Other activities include a Zoo train, a carousel, elephant and pony rides, a horse-drawn trolley, and an outdoor maze. The bottom line is that if you can't find an activity in which to participate at the Indianapolis Zoo, then you better check your pulse - you just might be dead!

Kahala Mandarin Oriental Hotel

5000 Kahala Ave
Honolulu, HI 96816
Phone: (808) 739-8888
Fax: (808) 739-8800
http://www.mohnl.com

Overview

Located on the southeastern shore of O'ahu, and just around the corner from Waikiki beach, the Kahala Mandarin Oriental Hotel is one of a large group of international hotels with a reputation for high quality service and style. The hotel reopened in 1996 after a $75 million face lift, and has returned to its former glory as one of O'ahu's most prestigious hotels. It originally opened in 1964 as the Kahala Hilton, and drew a large crowd of famous dignitaries including presidents Carter, Regan, Bush, and Clinton, Queen Elizabeth II, and the late Emperor Hirohito of Japan. Television and movie stars like Lucille Ball, John Wayne, Frank Sinatra, Bette Midler, and Tom Selleck were also seen at the resort, earning it the nickname "Kahollywood".

Whether or not the new Mandarin Oriental will become as popular as its predecessor remains to be seen, but it is certainly equipped to do so. Through the wisdom of the Mandarin's management, it has retained a large portion of the original staff, keeping one of O'ahu's most experienced resort teams intact despite intense recruiting efforts by competing establishments. One of the resort's greatest assets is its secluded location in an exclusive residential district, which is very attractive to famous and prominent guests. The hotel's grounds feature panoramic views of exotic gardens, a large cascading waterfall, ocean view restaurants, and a dolphin lagoon.

The hotel's 26,000 sq. foot natural lagoon is home to three Atlantic bottlenose dolphins. An educational feeding program takes place daily at 11 AM, 2 PM, and 4 PM. You'll also find tropical fish, sea turtles, exotic gardens and a waterfall.

Employment, Intern, and Volunteer Opportunities

Animal care and training for the Kahala's animals is handled by experienced personnel from nearby Sea Life Park (see page 124), and there are no internships or volunteer positions available from the hotel itself.

Research

It is not yet certain whether the Kahala's format can possibly allow for outside researchers, but there is not currently any research being done with the hotel's animals.

Other Programs

The Kahala offers visitors educational information about its dolphins, and often receives groups of schoolchildren on educational field trips.

Long Marine Lab

100 Shaffer Rd.
Santa Cruz, CA 95060
Phone: (408) 459-2883
Fax: (408) 459-3383
http://natsci.ucsc.edu/research/ims/LML.html

Long Marine Lab is, for undergraduate students, one of the best ways to get involved in marine mammal care and training. The Laboratory, a part of the University of California, Santa Cruz, has a long history in marine mammal research, and has sent many of its volunteer staff members into careers in cetacean care.

Overview & Research

The Joseph M. Long Marine Laboratory was dedicated to the founder of Long's Drugs (a great supporter of the University) and rests on 40 acres of some of the most gorgeous coastline in California. (Only in California would a marine laboratory be named after a drug dealer.) Overlooking the north end of Monterey Bay, the lab is a stone's throw from the Monterey Bay Aquarium and the MBA research Institute, Moss Landing Marine Labs, Hopkins Marine Station (of Stanford University), and of course, the diverse and species-rich waters of Monterey Bay itself.

The Laboratory is a part of the UCSC's Department of Marine Science, which is easily one of the top Marine Science programs in the United States. There are four main projects in progress at LML, three of which involve pinnipeds. The first two, which are done in concert with one another, are run by Dan Costa of the UCSC Physiology department and Jim Harvey from Moss Landing and San Jose University. These investigators are training sea lions to carry camera equipment in an open ocean setting, track down large whales, and - get this - TAG the whale with a transmitter while the camera records the whales' movements. It's pretty exciting stuff, and the open-ocean phase of the work should have just started by the time you read this. The researchers are also taking advantage of the training sessions to gather physiological data on free-swimming sea lions (something which was previously unavailable.) Ron Schusterman, a professor at UCSC, also maintains some pinnipeds at LML for his ongoing psychobiology programs at UCSC.

The fourth project underway at LML is an attempt to map out the physiological dynamics of Atlantic bottlenose dolphins' movement through water. Terrie Williams, another professor at UCSC and the project's investigator, is attempting to decipher both the mechanics of dolphin movement as well as the changing energy requirements of dolphins' movements through various stages of the animals' lifetime (i.e., nursing, hunting, living in cold or warm waters, etc.) The project is being carried out with the assistance of two ex-navy bottlenose dolphins, which are currently LML's only resident cetaceans.

Employment, Intern, and Volunteer Opportunities

Long Marine Lab is an all-volunteer outfit. The two personnel who are in charge of the facility are paid, but everyone else works for free. It isn't easy to get a job there, and the demands are great. The initial commitment for a new staff member is one year's worth of work at fifteen hours per week, with only a couple of weeks' vacation time in the summer. As with any entry-level position, the tasks are arduous - lots of cleaning, food preparation, water quality analysis, record keeping, and more cleaning. Newcomers are given a three-month trial period before being completely accepted to the program. Naturally, opportunities to assist in animal work increase over time. After one to two years of work and study, capable individuals progress to training animals themselves, at which point their time commitment increases to between twenty and twenty-five hours a week.

LML announces open positions once a year, usually between September and December, and usually through local flyers at local campuses. Roughly one to four hundred applicants arrive on the designated day for a mass orientation. Typically four or five positions open each year, so competition kind of goes through fierce and comes out the other side. LML staff use the opportunity to describe exactly how harsh and demanding the jobs are, and to take applications from everyone. Amazingly enough, every qualified applicant is called back, regardless of how many there are. No real emphasis is placed on prior experience, although any evidence of sustained commitment to animal care wouldn't hurt. Applicants who can offer more than one year's commitment are sought out - usually freshmen or sophomores at UCSC, though local non-students have been accepted as well. Most staff members stay with the program for two years, some for four, and one or two have stuck around even longer.

It is easy to weave the LML experience into an undergraduate education. College credit can be given for the staff's required lecture series, the job can qualify for work/study credit, and projects at the lab can be used to form senior theses. A freshman just entering UCSC with a strong desire to enter a career in cetacean care and research would be in the strongest position imaginable to take full advantage of the LML experience. Such a student could stick with the program all four years and graduate with a degree, first hand experience in marine mammal research, a fistful of contacts in the field, and would already be an accomplished trainer! It's a lot of work, but for those few individuals who have had a true calling to this field, LML could easily be the fast track to success.

It's also worth noting that Santa Cruz is a great party town.

Marine Animal Productions

(Marinelife Oceanarium)
P.O. Box 4078
Gulfport, MS 39502
Phone: (601) 864-2511
Fax: (601) 863-3673

Marinelife Oceanarium may look like an ordinary, run-of-the-mill marine park from the outside, but the facility is the headquarters of a much larger organization. Marine Animal Productions runs marine mammal shows in half a dozen parks and zoos across the country, and several others in foreign countries. Marinelife Oceanarium serves two purposes. It is a public display facility with all the exhibits and gift shops one comes to expect of such a place, but it is also a training ground and staging area for MAP's other locations.

Overview

The Oceanarium opened in 1956, making it one of the oldest marine parks in the world. It still sports one of the world's largest covered habitats, and houses dozens of bottlenose dolphins and pinnipeds. The park is showing its age, and many of its facilities are in need of repair and upgrade. MAP and many local businesses were disappointed in 1996 to see Donald Trump back out of a development plan for a casino in Gulfport. This would have meant a $10 million cash infusion for the oceanarium. Since then, other developers expressed an interest in the facility, making Marinelife Oceanarium's future uncertain, but hopeful.

Employment, Intern, and Volunteer Opportunities

There are just over a dozen volunteers at the Oceanarium, and the satellite facilities have been known to take on volunteers as well, though on a more limited basis. Volunteers are involved with food preparation and cleaning, and may occasionally assist trainers in animal sessions.

There are about nine trainers assigned to the Marinelife Oceanarium, three more in the Oklahoma City zoo, two at Knott's Berry Farm in California, and one in Hershey Park in Pennsylvania, all of whom work with bottlenose dolphins. (There are a number of other MAP teams elsewhere which deal with pinnipeds.) The Hershey Park crew is only active in the summer months; at other times the trainer and animals can be found at Marinelife. Staff members work with pinnipeds and cetaceans on a daily basis, and are often called upon to maintain water quality, narrate shows, and interact with visitors.

The staff members at satellite facilities have been known to take on part-time assistants. These are usually locals with limited animal experience, such as volunteer or intern time at other zoos and aquariums. Applicants for permanent positions must have college degree, though no particular degree is specified. (Psychology or biological sciences preferred.) MAP looks for individuals with volunteer or intern experience from other facilities. Interviews may be conducted at satellite facilities, if appropriate. Because of its lack of volunteer and intern programs, MAP is a good facility to attempt to get an interview from based solely on your resume. Approaching an MAP staff member at an IMATA conference might also result in an interview.

Contact information for the satellite facilities with cetaceans is given below:

Hershey Park
100 Hershey Park Drive
Hershey, PA 17033
Phone: (717) 534-3349

Knott's Berry Farm
Pacific Pavilion
8039 Beach Blvd.
Buena Park, CA 90620
Phone: (714) 220-5379
Fax: (714) 220-5380

Oklahoma City Zoo
Aquaticus
2101 NE 50
Oklahoma City, OK 73111
Phone: (405) 424-3344
Fax: (405) 425-0207

Research

Marinelife Oceanarium participated in many outside research programs in the past. Most recent projects include immunology studies, morphology, and sound reception. Proposals are given full consideration, but shouldn't interfere too greatly with set schedules. Projects requiring observations or blood analysis would be more likely to be accepted than cognition, echolocation, or other time-consuming studies.

Marineland of Canada

7657 Portage Rd.
Niagara Falls, Ontario
L2E 6X8
CANADA
Phone: (905) 356-8250
Fax: (905) 356-6305

Overview

Marineland of Canada, one of only three cetacean facilities in all of Canada, is a large park with marine shows, amusement rides, and animal displays. Just across the border from the American Niagara Falls, Marineland is a popular destination for families, with an emphasis on entertainment.

Employment, Intern, and Volunteer Opportunities

Of the two marine mammal facilities in the Niagara Falls area, one is a better choice than the other for American job applicants, and this ain't it. As with the other two Canadian dolphin habitats, it is VERY difficult to get a permit to work at Marineland of Canada. The only way to do it is to have Marineland of Canada go to bat for you with the Canadian government. The only way you'll get them to do that is to be very, very talented at marine mammal training, in which case, you don't need this book. (I'm not refunding your money, though).

If you're a Canuck, though, read on. Orcas are a huge attraction at Marineland. There are seven of them, with more possibly on the way with the completion of a new four million-gallon habitat which should be finished by the time you read this. Marineland trainers also work with more than a dozen sea lions, three bottlenose dolphins, two grey seals, and a partridge in a pear tree (currently on sabbatical). Marineland's staff members probably get a bit more exposure to actual animal training than many of their colleagues since their park is closed nearly half of the year, removing the need to accommodate guests. The best way into the Marine Mammal Department is probably through the two summer "trainer assistant" positions which usually run from May through August. It's a 40 hour/week job at a fairly low rate, but they often take younger applicants (17 and up) with little or no experience.

There are no volunteer positions at Marineland. That isn't unusual in foreign countries. I've found that we Americans are pretty much the only people dumb enough to work for free. In fact, it's illegal in many countries, like Australia, for instance.

Any applicants for a permanent position should have some animal care experience and a degree, while not necessary, would be helpful if it is in a scientific discipline. A SCUBA certification is helpful. A decent stage presence is necessary as well, as the job requires energy and enthusiasm before the public. (Lance Henriksen need not apply.)

Research

Although Marineland has conducted few research programs in the past, the management has plans to do more in the near future. The new cetacean habitat will serve as a breeding and educational facility, and will more readily accommodate a research climate.

Marineland of Florida

9507 Ocean Shore Blvd.
Marineland, FL 32086
Phone: (904) 471-1111
Fax: (904) 461-0156

Overview

Marineland was the first facility in North America designed to house dolphins, having opened in 1938 under the name "Marine Studios". The term "oceanarium" was created to describe the facility, and it pioneered many dolphin care techniques which are still in use today. In fact, visitors are given a kind of double exposure when visiting Marineland, as the facility is almost a museum for its pioneering solutions in exhibitry and husbandry. Water filtration, algae control, and a host of medical, engineering, and training methods had to be adapted from completely unrelated fields or created from scratch to get the nation's first marine park up and running.

The park now houses nearly two dozen bottlenose dolphins, as well as sea lions, penguins, sharks, and over a hundred other species of fish and invertebrates. It is a family-oriented tourist attraction, with an emphasis on entertainment.

Employment, Intern, and Volunteer Opportunities

Marineland has a small staff with very little turnover. There are only four full-time animal care technicians, which are assisted by two part-time workers. A dolphin-watch position opens up more frequently, though this is probably due to the relative inactivity the job demands.

Marineland is not currently accepting interns or volunteers.

Research

Early research projects conducted at Marineland by government and independent scientists led to advances in the treatment of birth defects and cancer, as well as shark repellent and sea survival kits for downed pilots in World War II. It was early research in cetacean bioacoustics at Marineland which revealed the dolphins' use of echolocation, or biological sonar. The University of Florida opened its C.V. Whitney Research Laboratory in conjunction with Marineland in 1974, which continues to investigate issues related to biochemistry, physiology, and coastal ecology.

Other Programs

Marineland is one of few facilities still accepting stranded cetaceans in need of rehabilitation. They respond to roughly 2-4 dozen strandings per year, though only a few are alive and can be brought back for recovery. As mentioned above, Marineland does not use volunteers to assist with the care of such animals.

Marine World Africa USA

Marine World Parkway
Vallejo, CA 94589-4006
Phone: (707) 644-4000
Fax: (707) 644-0241

Overview

Marine World Africa USA has a long and sometimes bumpy history stretching back nearly thirty years. The product of a union between two separate animal parks, MWAUSA has moved and changed hands several times, including a transfer of ownership in 1996 to the City of Vallejo.

The park is one of the largest of its kind. It encompasses 160 acres, employs over one thousand people during the summer, and hosts over one million guests each year. MWAUSA features a large number of animal shows involving orcas, bottlenose dolphins, sea lions, water ski teams, "aquatic" tigers, and numerous others. The shows are designed to be entertaining, though exhibits like the "Wildlife Theater" give children a chance to learn up close about cheetahs, binturongs, squirrel monkeys, and other unusual animals. In the past, MWAUSA has been more progressive than many other "theme park" facilities in the areas of research and education.

Employment, Intern, and Volunteer Opportunities

MWAUSA does utilize volunteers in a variety of support roles, but they are not allowed to interact with the animals. Internship opportunities at the park are similarly spartan. The park is a host facility for Moorpark College internships but there is no formal internship program for other students. MWAUSA cites concerns about potential insurance hassles, so if an enterprising student could secure insurance through his/her college, MWAUSA might consider taking them on as an intern. (Though I get the feeling it would be an uphill battle to do so.)

The park hires about four seasonal employees in early May to work through September. Nearly all seasonal employees come from moorpark college, though occasionally interns from other marine mammal facilities are hired. Applications should be submitted by february.

There are just over a dozen permanent marine mammal trainers at MWAUSA, divided into a pinniped group and a cetacean group.

Positions seem to open up once every other year on average. New positions are often filled by prior seasonal employees, though one or two employees were hired based on volunteer experience at other facilities. MWAUSA sometimes sets up interviews on the spot at IMATA conferences. When staff vacancies occur, MWAUSA occasionally fills them with relatively inexperienced candidates.

Research

MWAUSA is committed to conservation, and to that end it conducts a wide variety of research projects, both in the wild and in captive settings. Dr. David Bain is a staff scientist and the park's Director of Killer Whale Research. Dr. Bain, continuing a fifteen-year study on orca vocalizations and social behavior, occasionally takes on graduate students to assist him. Other work at MWAUSA focuses on cetacean cognition, physiology, and rehabilitation tactics for stranded animals. Individuals interested in assisting with current projects, or graduate-level researchers with proposals for new studies, should contact the General Manager at (707) 644-4000.

Other Programs

Like many other groups with permanent cetacean residents, MWAUSA had gotten out of the stranded cetacean rehab business, due to the risk of morbillivirus infection. the park recently formed a partnership with The Marine Mammal Center in the Marin Headlands, California, to allow the program to be reinstated. Cetaceans in need of rehabilitation are held at TMMC until the possibility of a morbillivirus infection can be ruled out through blood tests. MWAUSA relies on its full-time staff to rehabilitate such animals.

Miami Seaquarium

4400 Rickenbacker Causeway
Miami, FL 33149
Phone: (305) 361-5705
Fax: (305) 361-6077

Overview

The Miami Seaquarium is one of the oldest marine mammal pavilions in the country, and is home to seals and sea lions, Pacific whitesided dolphins, and orcas. Featured exhibits include a shark habitat, a coral reef tank, and the "Faces of the Rainforest", which displays birds, reptiles, and other tropical rainforest inhabitants.

The Seaquarium is a family-oriented marine park with an emphasis on entertainment. The facilities have been updated, the education department has been bolstered, and the life support, animal care, and training staffs are exceptional. The aquarium is invaluable to the manatee population of southern Florida as the premiere rescue and rehab center for these endangered animals.

Employment, Intern, and Volunteer Opportunities

At present there are no volunteers or interns in the training department, though plans are being discussed to open that possibility in the future.

There are 20 full-time staff members in the training department. Each member is assigned to one animal group (bottlenose dolphins, orcas & Pacific whitesided dolphins, etc.) but the staff is often rotated to keep everyone constantly learning new things. The Seaquarium staff places emphasis on the team approach. Everyone's input and skills are valued, and individuals are permitted to progress at their own rate, giving them a chance to excel in those areas where their aptitude is greatest.

Positions open up once or twice each year. The Seaquarium is very selective as cetacean facilities go. College degrees are preferred, though anything related to psychology or the life sciences is acceptable. A SCUBA certification is not mandatory, but a good idea. Applicants must take a fairly strenuous swim test.

Almost all successful applicants had experience in other facilities, whether as an intern, volunteer, or staff member. Most important of all, the Seaquarium staff won't take a chance on someone until they are confident that person can work well with the rest of the team. This can often mean that an applicant will be passed up for a few positions until the staff has come to know his or her personality and seen a fair degree of persistence. If you are determined to work here, go and visit them whether they are hiring or not, and continue to follow up on your application.

It might also be worth noting that the Miami Seaquarium maintains a staff of pool maintenance workers who routinely enter the animal habitats in SCUBA gear. Naturally, extensive SCUBA experience is a must for these positions, which involve maintenance on animal habitats as well as water quality analysis and control. Laboratory experience would also be helpful.

Research

The Seaquarium conducts studies on manatees, dolphins, and turtles in areas like immunology, pathology, and microbiology. Researchers from local schools like the University of Miami and Eckerd College, as well as from the NMFS Miami Lab often enlist the Seaquarium's aid in conducting a wide range of research projects.

Other Programs

The Seaquarium is an active member of the Southeast Marine Mammal Stranding Network, and assists the National Marine Fisheries Service in responding to numerous cetacean strandings each year.

Educational programs offered at the Seaquarium include one-say mini camps, spring and summer camps, field trips, guided tours, teacher workshops, and outreach programs. More than 90,000 students take advantage of the park's programs each year.

Minnesota Zoo

13000 Zoo Blvd.
Apple Valley, MN 55124
Phone: (612) 431-9200
Fax: (612) 431-9300
http://www.mnzoo.com
zooinfo@wolf.mnzoo.com

Minnesota (The Mall State) may be the last place you'd expect to find cetaceans, but they are there all the same. The Minnesota Zoo, which is just south of Minneapolis/St. Paul, houses six bottlenose dolphins in a one million-gallon indoor exhibit. The dolphin habitat is part of Discovery Bay, a brand new multi-million dollar addition which features a host of marinelife and exhibitry.

Overview

The Minnesota Zoo opened in 1978. Large, foresty, and spacious, the zoo offers a series of animal "trails", each exposing visitors to groups of animals and activities related to different biomes. The zoo's animal collection consists of over 2,300 animals representing species from all over the globe, including 15 on the endangered species list. The zoo participates in the Species Survival Plan program, and places a heavy emphasis on conservation and "strengthening the bond between people and the living earth." Overshadowed in attendance only by that most garish of temples to consumerism, the Mall of America, the zoo is one of Minneapolis' most popular tourist destinations. In addition to the dolphin habitat, Discovery Bay also features a shark reef, tide pool, and an estuary exhibit. Although the zoo's cetacean populace consists solely of bottlenose dolphins, the design of the new facility has opened the possibility of introducing other species at a future date.

Employment, Intern, and Volunteer Opportunities

The cetacean staff is small, consisting of six full-time employees. Obtaining employment at the zoo can be difficult due to its low turnover rate. Applicants for full-time positions must have bachelor's degree in an appropriate field, a year's worth of experience in cetacean care and training, and be SCUBA certified. Acceptable fields of study include the biological sciences and Zoology. An additional year of experience can be substituted for a college degree. Intern experience at the zoo can count toward the experience requirement, and many trainers have come from the intern ranks. Applicants which meet hiring criteria take a written essay test. Those who pass the written test are placed on an eligibility list, where their name will sit until a position is open.

Internships are offered to college students from all over the country. The internships are unpaid, and generally last about three months. Internships must be done for college credit through the student's home school. Students may apply to complete more than one internship back-to-back.

(Photo provided by the Minnesota Zoo)

The Minnesota Zoo features a brand-new, state-or-the-art cetacean habitat.

Research

Like many display facilities, Minnesota Zoo doesn't have the resources for a full-blown cetacean research program. They do participate in studies conducted by other facilities or outside researchers. Recent studies include cetacean birth control and EKG equipment, and the zoo is considering proposals for future studies.

Other Programs

The Minnesota Zoo offers a slew of educational programs for youngsters, teens, and adults. Group activities for classes of kids, summer camps, teacher seminars, and adult classes focus on a rainbow of topics from natural history and biology to environmental issues to zoo design and management. Occasionally a college credit or two can be earned in cooperation with nearby Mankato State University, the University of Minnesota, or Hamline University. Very few classes have involved the zoo's cetaceans in the recent past, but offerings vary from year to year. (Although it offers no regular cetacean classes, the University of Minnesota would probably be the most receptive of the three to a proposal of some kind.)

The zoo's cetacean staff also participates in "Dolphin Dark to Dawn" sleepover programs. These are offered to families or special groups, and features a number of fun and educational activities. Contact the zoo's education department for current schedules of zoo programs.

Mirage Dolphin Habitat

P.O. Box 7777
Las Vegas, NV 89177
Phone: (702) 791-7588
Fax: (702) 792-7684

Many of you may be skeptical of a major Las Vegas strip resort maintaining a dolphin habitat (Lord knows I was.) It's easy to assume the hotel is simply interested in luring more patrons and stuffing its pockets with just that much more cash. After all, that's what Las Vegas is all about, right?

Certainly the Mirage's dolphin habitat was fashioned as an attraction, designed to bolster the overall enjoyment of the guests' experiences, and as with any licensed display facility in the United States, the design of the habitat and the actions of the staff place the needs of the animals above all. The fact that the habitat was designed first and foremost as an educational experience for its guests was a very pleasant surprise. The Mirage has made more effective use of a cetacean display facility as an educational tool than many other organizations I've seen. In doing so it has forgone a significant amount of income it might have had under a glitzier, less responsible format. This ecologically conscious attitude is reflected elsewhere in the resort's policies, such as its aggressive recycling program and its bans on fur sales and the use of "dolphin safe" tuna obtained by questionable means.

Overview

The Mirage habitat consists of four interconnected pools holding 2.5 million gallons of water, making it one of the largest in the country. The habitat is outdoors, surrounded by Las Vegas' usually gorgeous weather. Artificial reefs, irregularly-shaped contours, and sandy bottoms help simulate a natural environment for the animals. The dolphins are Atlantic bottlenose dolphins, all captive-born or relocated from other facilities. There are no scheduled dolphin shows for the public to watch. Rather, guests are given a guided tour of the facility and its animals by an educational docent, and are then invited to roam the habitat at will as long as they like.

Employment, Intern, and Volunteer Opportunities

The Mirage is probably in the upper ten percent of facilities with respect to how difficult it is to get a job. Let me paint you a picture: Do you have more than six months' experience working or volunteering with animals? Have you taken college classes in the biological sciences? Do you already have a working knowledge of basic marine mammal natural history? Can you demonstrate an unwavering determination to work with animals, whatever it takes? If you can't answer all of these questions with a resounding "yes",

(Photo provided by the Mirage)

One of several calves successfully born and raised at the Mirage Dolphin Lagoon.

then I don't have anything terribly pleasant to tell you about your chances of landing a job at the Mirage. They are very picky, and can afford to be with something upwards of one thousand applicants per position. There are only about eight positions total, by the way, and the openings are rarer than at many other facilities. I'm not saying it's impossible to get hired there; it's just very difficult.

The Mirage has no volunteer or intern programs. Paid docents are used to conduct tours for the public and answer questions. The docents occasionally assist in the animals' training sessions, and once in a while one even gets hired on as a trainer. The docent positions are considerably easier to secure.

Other Programs

The Mirage has a host of educational programs for students in pre-school all the way up through high school. The programs are well-suited for each age group, and do a good job of remaining informative while keeping the students' interest. Seminars on a variety of marine-related subjects are offered to college students throughout the year, and the Mirage runs a number of lectures and special events through its community outreach program.

Mystic Marinelife Aquarium

55 Coogan Blvd.
Mystic, CT 06355-1997
Phone: (860) 572-5955
Fax: (860) 572-5969

Mystic Marinelife Aquarium

Mystic is a pretty small town, and by "pretty small" I mean it has to borrow cops from its neighbors since it has none of its own, and the fire department's most exciting event in recent history was helping me break into the local Amtrak station to retrieve the duffel bag I'd accidentally locked inside. (I guess it wasn't really a "station" so much as a large wooden crate with a door near the tracks.) The fire chief himself got all wide-eyed and psyched up to finally be dealing with a crisis of some kind. The switchboard operator was told to raise Amtrak security, and a couple of firemen were sent to contact the owner of the building while the chief grabbed a crowbar and headed off to check the feasibility of forced entry (which is the sort of thing you can do if you're the chief.) The whole thing was kind of embarrassing.

We never did get the bag out that night, so the fire chief put me up in the firehouse (alone) with the understanding that I wouldn't steal any of the fire trucks. Coming from downtown Chicago, the trusting behavior of small-town America never ceases to astound me. The firehouse turned out to be more comfortable than my loft apartment back home! Huge plush couches, a new pool table, a washer and dryer, a complete kitchen, an enormous TV, and the biggest entertainment center I've ever seen form the nucleus of Mystic's fire-fighting equipment. All in all, it was the most modern firehouse I've ever stayed in, which just goes to show you that being small has nothing to do with the quality of one's resources. This is just as evident in the Mystic Marinelife Aquarium, which is one of the fastest growing aquariums in the nation.

Overview

The Mystic Marinelife aquarium is a large public aquarium with a focus on education and conservation. Since its inception in 1973 it has housed beluga whales, bottlenose dolphins, marine birds, pinnipeds, and thousands of fish and other aquatic life. MMA's pinniped collection is the largest and most diverse of any non-profit zoological institution in the United States. The aquarium's animal husbandry staff is divided into three sections: Seal Island, Fish and Invertebrates, and the Marine Theater. The aquarium has internships available in all three areas, as well as in a number of non-husbandry departments. Mystic also maintains an active volunteer program, and offers a multitude of educational opportunities for students of all ages.

MMA recently opened its brand-new Aquatic Animal Study Center, a facility designed for the research and care of off-exhibit cetaceans. By 1999 the Center will also be an effective rehabilitation center for stranded dolphins and whales. At the same time, work continues on the Alaska Coast exhibit, an outdoor beluga whale habitat which should also be completed before the Millennium. MMA's statistics gurus predict an increase in attendance to a record-setting 1.2 million visitors per year.

Employment, Intern, and Volunteer Opportunities

The Marine Theater group consists of about seven full time aquarists who work with the beluga whales and dolphins. All staff members are expected to learn to narrate public presentations. Narrations are educational in nature and focus on the natural history and biology of cetaceans. The Marine Theater staff is small and fills relatively few vacancies. Nevertheless, many staff members are ex-volunteers or interns. Staff members from Seal Island often find their way into the Theater as well. Hiring criteria at Mystic may change from time to time based on the current staff's strengths and weaknesses. MMA does not actually require a college degree, but as with many facilities, it's a strike against you if you don't.

The intern program at MMA offers internships in twelve different departments including the three Husbandry groups. Internships are designed to fit the needs of the student while remaining consistent MMA's mission and policies. Each intern devises a research project, either alone or with the assistance of staff members, which serves as the intern's primary goal. While other MMA internships may last from four weeks to three months, and entail anywhere from twelve to thirty-five hours each week, the Husbandry internships always last for a full semester, and generally require the greatest hourly time commitment. Marine Theater interns do a lot of cleaning and food prep, but also become involved in the animals' training sessions. As with most internships, superior performance may garner greater opportunities, and some interns have been asked to remain with the aquarium beyond the period specified in their agreement. Applicants for all internships must be able to receive college credit for the experience, but the program is open to both undergraduates and grad students. Surprisingly enough, during some semesters there has been little or no competition for the Marine Theater's internships (two interns can be accommodated each semester.)

The volunteer program at MMA is fairly typical. Volunteers in the husbandry areas are expected to do one shift per week on the same day, and are assigned with a partner to either a morning or afternoon shift. Weekend volunteers come in every other week. Volunteers must be 18 years of age to work in the Husbandry areas; 15 years of age in other departments. Marine Theater volunteers clean and prepare food, and enjoy a fair amount of contact with the animals.

Research

MMA operates a research department and is equipped with a comprehensive laboratory, many of the aquarium's research projects involve cetaceans. Field studies on Arctic belugas and South Carolina dolphins have been conducted by MMA staff members for several years. MMA has generated more than 100 articles published in a variety of scientific journals. New projects are always being considered, and proposals are reviewed by the Aquarium's department of Research and Veterinary Services.

Other Programs

MMA is extremely active in cetacean stranding response, and was a founding member of the Northeast Regional Stranding Network. MMA staff respond to both live and dead strandings throughout the Connecticut and and Rhode Island coastlines. Once the Aquarium's new facilities are completed, MMA will also assume an even more active role in bringing stranded animals back to health.

MMA has one of the most extensive educational programs on the east coast, reaching out to more than 70,000 school children each year. For classes of preschool, Kindergarten, first through third-graders, and fourth through sixth-graders, the aquarium can provide a number of special presentations on subjects ranging from jellyfish, tide pools, food chains, and algae to endangered marine mammals, penguins, and the use of camouflage in the sea. Each of these programs runs for about 45 minutes and can accommodate twenty or thirty students, depending on the age group. A similar assortment are available for grades seven through twelve. More involved programs are also offered, lasting upwards of two hours and consisting of hands- on classroom work, field studies, or in-school lectures. Teaching aids, teacher workshops, and orientations are available as well, and can often be tailored to meet an educator's needs.

Offered to individual members are an assortment of seminars and classes on an equally diverse range of marine topics, and for a substantially lower cost. Member's programs are offered for children, adolescents, teenagers, and adults. Most programs cost five or ten dollars, though overnight events are a bit more.

Mystic is one of very few aquariums to regularly offer college level classes. Conducted through the University of Connecticut, several undergraduate-level marine courses are offered each year, often including "Marine Mammal Science", an introductory marine mammology study which offers exposure to current research and husbandry practices. For more information on these or any of the above programs, contact the Education Department at extension 204.

National Aquarium in Baltimore

501 E. Pratt St. Pier 3
Baltimore, MD 21202
Phone: (410) 576-3850
Fax: (410) 576-1080

Of all the architectural structures in Baltimore's touristy Inner Harbor area, the National Aquarium is definitely the weirdest thing going. The style of architect Peter Chermayeff, who also designed the Tennessee Aquarium, the New England Aquarium, Japan's Osaka Ring of Fire Aquarium, and now a beast of a structure called the Oceans Pavilion for Expo '98 in Lisbon, has been referred to as "striking", "stunning", and "magnificent". It certainly catches your eye.

Overview

In 1979, the U.S. House and Senate voted unanimously to designate the multimillion-dollar aquarium being built in Baltimore, MD, as the "National Aquarium". Nearly twenty years and a Marine Mammal Pavilion later, the National Aquarium in Baltimore is still one of the nation's premiere public aquariums, and welcomes more than 1.5 million guests annually.

The inside of the aquarium is an exciting assortment of innovative displays, galleries, and atriums. Over 5000 specimens representing more than 500 species are on display throughout. The aquarium features an enormous open-air shark and ray tank, some funky bubble-tower things, and a gorgeous four-story circular reef tank which surrounds a spiral viewing ramp. At the top of the structure is a fascinating reproduction of a South American rain forest enclosed in a glass pyramid. (I have particularly fond memories of literally bumping into one of the pyramid's incredibly mellow two-toed sloths. It slept through the entire encounter.)

The Aquarium's Marine Mammal Pavilion, added in 1990, is a completely separate building designed to compliment the first, and is home to about a half-dozen bottlenose dolphins. The Pavilion features scheduled performances with the dolphins - well scripted and education-rich presentations which make very effective use of audio/visual materials and transparent habitat walls. They are bit showy, but nevertheless very well done. Roughly a dozen trainers work with the dolphins, as well as the harbor and grey seals in the main building's outdoor seal pool.

Employment, Intern, and Volunteer Opportunities

While NAIB makes use of volunteers, the Marine Mammal Department does not. Volunteers may choose to work as an exhibit guide, information desk attendant, or as an office, aviculture, horticulture, aquaculture, herpetology, library, computer, or aquarist assistant. Or, if none of that appeals to you, you can also choose to don the garb of the aquarium's giant puffin mascot and interact with the visitors. (For free, no less!)

Volunteers who have served for a minimum of six months are given the opportunity to apply to the Marine Animal Rescue Program. NAIB is very involved in stranding response along the Maryland coast, and often rehabilitates small cetaceans and pinnipeds in its veterinary pool. The rescue program maintains a list of rescue volunteers who are on call 24 hours a day to assist in rescues, observations, feedings, and habitat maintenance and cleaning. Rescue volunteers may choose to continue in their first volunteer position, but are requested to commit to one year of on-call duty, and are expected to participate in twelve hours of outreach activity (parades, festivals, newsletters, etc.)

There are quite a few intern opportunities at NAIB. Nearly every department uses them (education, exhibits, publications, etc.) and the Marine Mammal department is no exception. College-level internships are open to any students who can receive credit for the internship, are in good standing with their school, and are majoring in a related field. Applicants should also have some knowledge and/or experience in this area. The internship is a 120 hour commitment, and is unpaid. Interns prepare food, clean and maintain habitats, record behavioral observations, feed seals under supervision, and occasionally may assist in animal training sessions. College internships are offered all year-round, usually four or five per semester.

High School internships are available, though they are fewer in number and more selective. These internships require a 75 hour commitment over the summer, and applicants are usually chosen in April and May. High School and college interns perform much the same tasks.

The Aquarium recently added "Aid" positions. Similar to summer internships, these paid full-time positions are open to non-students. Aids share the same tasks as interns, but are often given greater responsibilities, and sometimes learn to narrate public shows.

When hiring for a full-time position, NAIB looks for individuals with some experience working or volunteering at another zoo or aquarium. A SCUBA certification would be helpful as well. With a staff of twelve, NAIB will most likely look to hire someone at least once a year, though as large organizations go, it has a relatively low turnover rate.

Research

The Aquarium's Marine Animal Rescue Program is conducting a population study of bottlenose dolphins in Chesapeake Bay, Delaware Bay, and adjacent areas of the Atlantic Ocean. The study used photo identification and data collection to map out the resident population, and to understand their habitat use and movement patterns. Aquarium volunteers are sometimes offered opportunities to get involved in this study on a case-by-case basis.

The nature of NAIB's marine mammal program precludes the possibility of most intensive research proposals, but should they be approached by a researcher with a proposal which would mesh well with their schedules, they will give it full consideration.

Other Programs

NAIB has an assortment of outreach programs available for local area schools, as well as informational packets and tip sheets for teachers bringing their students to the aquarium. There are no on-site classes offered at the facility, but there is currently a Master's degree program being offered in conjunction with the University of Maryland at Baltimore County. "The Biology of Aquatic Organisms in Artificial Environments" is designed for biology majors who wish to specialize in the study of aquatic organisms in an artificial environment. Contact the UMBC Department of Biological Sciences at (410) 455-3669.

Oregon Coast Aquarium

2820 SE Ferry Slip Rd
Newport OR 97365
Phone: (541) 867-3474
Fax: (541) 867-6846

The Oregon Coast Aquarium is the only cetacean display facility in the world to foster cooperation between cetacean care specialists and animal rights activists. The possibility of relocating Keiko to a large, natural seawater habitat in preparation for possible release became a reality through the efforts of organizations like Earth Island Institute, which is the leading member of the Free Willy-Keiko Foundation, and a traditional opponent of captive cetaceans. To ensure Keiko's good health, however, they found themselves enlisting the aid of experienced cetacean care personnel from such facilities as the Point Defiance Zoo and Aquarium, the Miami Seaquarium, and Marineland of Canada. Although Earth Island Institute and the other members of the foundation are still opposed to captive cetaceans, they are now working side-by-side with former adversaries whose primary motivation, Keiko's well-being, is the same as their own.

That's a step in the right direction.

Overview

Located in Newport about an hour's drive from Portland, the Oregon Coast Aquarium lies right on highway 101, smack dab in the middle of some of the United States' most beautiful coastline. Though not one of the biggest, Oregon Coast is one of the more beautiful Aquariums I've seen. The designers and architects did a wonderful job of harmonizing the outdoor exhibits to the point where visitors get the impression of wandering here and there through the real Oregon coastline. The aquarium's theme follows the path of a raindrop landing inland, running downriver, through estuaries, and into the open ocean. Phase one of this project, the inland exhibits, was completed prior to Keiko's arrival, but phase three, the open ocean exhibit, was pushed ahead of schedule to accommodate him. Phase two, the river and estuary exhibits, should be nearing completion by the time you read this. The

Aquarium features sea otters, seals, sea lions, seabirds, an octopus exhibit, and some of the best jelly exhibits in the country.

For those of you hermits who missed the whole Keiko story, here's a quick rundown. In 1979 the whale was captured at the age of one or two near Iceland and placed in an aquarium near Reykjavik. Three years later he was sold to Marineland of Canada, and three years after that he went to Reino Aventura in Mexico City. He stayed in Mexico for seven years before starring in "Free Willy" in 1993. By 1995 the Free Willy-Keiko Foundation was formed by Earth Island Institute, Warner Brothers, New Regency Productions, and The Mystery Donor. (Somebody made a huge anonymous donation.) By late '96 Keiko's Mexican owners agreed to hand him over free of charge, and the Foundation arranged to have Keiko brought to a new two million gallon open ocean exhibit at the Oregon Coast aquarium.

Will Keiko be released to the wild? It's impossible to say right now, but Keiko has indeed made tremendous progress. He arrived in Oregon diseased, underweight, and lethargic. After more than a year of rehabilitation the whale has gained about 2,000 pounds, is significantly more active, and his skin condition has cleared up nicely. Any animal to be released into the wild must, at the very least, overcome its dependency on humans. Keiko, while having come so far in his recovery, still requires human contact for food and medical observation, and won't be in a position to learn any real self-sufficiency for quite a while.

Employment, Intern, and Volunteer Opportunities

There are volunteers at the Oregon Coast Aquarium (over two hundred), and some of them even work with the aquarium's seals and sea lions, but opportunities to assist with Keiko are extremely rare, and mostly reserved for more experienced volunteers. New volunteers assist in cleaning, food preparation, and record keeping. Volunteers are given much opportunity to learn about training and animal care. They are given 24 hours of training in natural history, marine biology, and teaching techniques, and are expected to progress during their time with the aquarium. Capable individuals are given the opportunity to work with and eventually train the aquarium's pinnipeds. Volunteers are only required to commit to one morning a week, though they are welcome to give more time if desired.

There are very few opportunities for paid positions. About half a dozen trainers work with Keiko, and many of them have a great deal of experience with orcas at other facilities. I don't expect to see much of a staff turnover rate over the next few years either, so the Oregon Coast Aquarium, while a wonderful facility, provides few opportunities for breaking into the field.

Summer internships lasting 12 weeks have been offered to college students in the past, though only as the Aquarium's budget allows. Most have come from local colleges, though anyone may apply. Biological science majors are preferred. The internships are paid, which is rare.

Research

Although the aquarium's goal is to rehabilitate Keiko, not to use him in experiments, the entire project itself is a test case for the feasibility of captive release programs for orcas. Indeed, the facility may one day be used for other cetaceans in similar circumstances.

Other Programs

Oregon Coast offers educational programs to more than 35,000 children each year. Over twenty programs are available to students from kindergarten through twelfth grade, including pre-visit activities, on-site classes, and auditorium lectures.

Pittsburgh Zoo

1 Hill Rd.
Pittsburgh, PA 15206
Phone: (800) 474-4966
Fax: (412) 665-3661
http://zoo.pgh.pa.us

Overview

The Pittsburgh Zoo is just about to celebrate its 100th birthday, and in that time the zoo has come a long way. Where animals once huddled alone in cramped sterile cages, they are now organized in ecologically appropriate social groups within naturalistic habitats, and the zoo continues to build on its progressive educational department.

Since time has shown that these animals do not fare well in groups, the zoo's single Amazon River dolphin is its only cetacean. The same keepers responsible for the dolphin also care for the zoo's penguins and several fish exhibits.

Employment, Intern, and Volunteer Opportunities

There are five keepers in the zoo's aquatic life department who are responsible for the zoo's aquarium. The keepers are union employees, which provides them a good degree of job security, which in turn yields a very low turnover rate.

There are unpaid internships available in the Aquatic Life Department, typically open to college students in the biological or veterinary sciences. Prior experience is not necessary, but helpful. This relatively new program is unpaid, and accepts from four to six interns at a time, all year round. Internships last for a minimum of three months, but the zoo would welcome longer stays. Interns are responsible for collaborating with their sponsoring professor in designing a project to be approved by the zoo and completed during the internship. Prior projects have included water quality analysis and preparing temporary exhibits.

There are no volunteer opportunities in the Aquatic Life department.

Research

The Pittsburgh Zoo is committed to furthering the worldwide shift of zoos' emphasis to conservation and education. To support this initiative the zoo conducts research on the behavior and environmental requirements of its animals, and encourages outside researchers to approach it with their proposals. (Grad student or above.) Interested parties should contact the zoo's Research Director, Dr. Langbauer, for information.

Other Programs

The Pittsburgh Zoo has been trying to maximize the use of its five-classroom Education Complex and its new Discovery Pavilion. From October through June the Education Department offers over twenty different programs focusing on topics like rainforest ecology issues, ocean mysteries, endangered species, and life on the African savanna.

The zoo offers a weekday and weekend program called "Wildlife Academy" which gives children and adults an opportunity to learn about almost any animal issues through lectures, video presentations, artifacts, stories, and behind-the-scenes tours.

In the summer the Zoo's Animal Adventures Day Camp opens for kids ages 3-13, giving children a chance to learn about animals and their environment through classroom activities, art projects, games and visits with the animals and their keepers. Contact the Education Department for details on any of these programs.

The zoo also offers Teacher Workshops which cover such topics as Ocean Overview, Nocturnal Creatures, and Tropical Rain Forests. These programs are designed to help educators gain knowledge of the natural world as well as innovative teaching strategies.

Point Defiance Zoo and Aquarium

5400 N Pearl St.
Tacoma, WA 98407-3218
Phone: (253) 591-5337
Fax: (253) 591-5448
http://www.pdza.org

Nestled in Tacoma's Point Defiance Park, and overlooking the breathtaking beauty of Dalco Passage, the Point Defiance Zoo and Aquarium is one of the most gorgeous places to work in the country. The park is a peninsula which juts out into the waters of Puget Sound, and wild orcas can sometimes be seen from the park shore. A 5-mile scenic drive and nearly 14 miles of footpaths wind their way through the vast park's gardens, gazebos, beaches, and historic exhibits. Although Tacoma itself isn't the most exciting city in the nation, or the safest, (in all of my travels it's the one place I got my butt thoroughly stomped) it's only a forty-five minute bus ride to Seattle, and the scenery makes up for many lost amenities. Bring an umbrella, though.

Overview

The PDZA is Tacoma's #2 tourist attraction (after Mt. Rainier) and continues to be recognized as one of the nation's top-rated zoos. More than 5,000 animals represent 350 species at the zoo, whose "Ring of Fire" themes feature species from Pacific Rim nations. The zoo is known for its groundbreaking work with elephants in protected contact, and the management of its Species Survival Program for the endangered red wolf.

There are really two aquariums at PDZA, an outdoor cold water section, and an indoor warm water facility, the latter of which features over forty species of sharks and numerous tropical fish. The zoo also features an indoor North Pacific bull kelp forest. The cold water facility houses the zoo's walrus, sea otters, penguins, puffins, and beluga whales. These habitats are a part of the zoo's "Rocky Shores" section, which also includes the red foxes and muskox. Animal care technicians in Rocky Shores are trained to work with all these animals, though individual staff members tend to specialize in certain areas.

Employment, Intern, and Volunteer Opportunities

PDZA hires experienced animal care technicians and marine mammal trainers to fill vacancies in its Rocky shores staff. Consequently, it is not a likely place for less experienced applicants to seek employment. Nevertheless, the Zoo has provided innumerable opportunities for entry-level experience through its volunteer and internship programs.

Internships are open to undergraduate and grad students. Preference is given to students in a related field of study. The PDZA psychological, sociological, and public speaking majors in addition to the traditional life sciences. Rocky Shores interns work in nearly every aspect of the department's daily operation, including food prep, cleaning, record keeping, and occasionally feeding. The zoo likes to recruit applicants with a rudimentary understanding of the rigors of animal care, so animal experience of any kind is very helpful. Internships are unpaid, and demand full-time participation. Two interns are brought on every season, for a total of eight per year. Interns complete a number of individual projects during their stay, one of which the intern designs. Projects can involve almost any aspect of marine mammal display and behavior, and may include behavioral studies, educational displays, and husbandry techniques, among other topics.

Volunteers are used throughout the zoo, including its animal areas. Volunteers get a wide variety of exposure to animal care techniques, and occasionally assist in animal sessions, but positions in Rocky Shores have grown fewer and farther between in recent years. Volunteers are asked to commit to four hours of work each week, usually on the same day.

Research

Like many of its fellow display facilities, most research proposals are beyond the Point Defiance Zoo and Aquarium's logistical capabilities, but they have participated in programs in the past, and would welcome proposals which might mesh well with their own schedules.

Other Programs

For grade school classes interested in Marine Mammals, sharks, or endangered species, the PDZA has a number of "focused field trip tours" and outreach programs. Overnight programs give children and adults a chance to get to know the place better, and specialized week-long programs may be designed for local school groups. "Keeper Camp" is a two-day program for students between thirteen and seventeen years old who have an interest in biology. Participants have an opportunity to work side-by-side with the animal care staff, prepare food, clean habitats, and get a general feel for what an animal care technician's day is like. The PDZA Education Department fields inquiries about any of the above programs.

Scientific Applications International Corporation

3990 Old Town Ave #105A
San Diego, CA 92110
Phone: (619) 294-8380
Fax: (619) 294-8795

Overview

Scientific Applications International Corporation is actually an enormous organization with over 17,000 employees worldwide. SAIC's Maritime Service Division is the prime contractor to the Navy for hiring animal care and training staff personnel under its Marine Mammal Services and Training contract. The staff of approximately 55 employees consists of Animal Trainers, Assistant Animal Trainers and Divers.

Volunteer, Intern and Employment Opportunities

SAIC is your best bet if you want to train marine mammals for the Navy, and in fact will hire entry level Assistant Animal Trainers with little or no specific experience in marine mammal training. The minimum qualifications for an SAIC job are a high school diploma, a SCUBA qualification from a nationally recognized diving organization like PADI, NAUI, or SSI, acceptable physical condition and health (able to lift 50 lbs.), and the ability to successfully complete a background investigation to gain a security clearance. Anything beyond those qualifications will improve your chances, such as volunteer, intern, or paid experience working with animals, or having a college degree in the natural sciences.

In the past, SAIC has hired Moorpark College graduates, ex-Navy mammal handlers and divers, and commercial divers for entry level positions within the animal training staff. Some of the diver positions are intended for dedicated diving services "to perform routine project support or underwater maintenance tasks." These positions are typically filled by ex-Navy divers or divers who have graduated from an accredited commercial diving school. See page 135 for more information on the Navy's marine mammal programs.

Sea Life Park

41-202 Kalanianaole Highway #7
Waimanalo, HI 96795
Phone: (808) 259-7933
Fax: (808) 259-7373
http://www.cyber-hawaii.com/vacation/corp/oaa/sealife/sealife.html

Overview

Sea Life Park is a family-oriented amusement park-style facility located on the southeastern tip of Oahu in the Hawaiian islands. It is has the highest profile of Hawaii's cetacean facilities. It's one of Oahu's main attractions, drawing thousands of visitors annually from the mainland, Australia, Canada, the Orient, and almost anyone else who drops into the isles. The Park is absolutely beautiful. Resting on the eastern shore and overlooking a handful of small islands, it's fun just to sit and compare the park's animals to the seabirds, fish, and occasionally even dolphins and whales which are visible to the east.

The park is a showcase for some of Hawaii's native aquatic wildlife, and also serves as a vehicle for conservation and education. Its many habitats, most of which are outdoors, feature seals, bottlenose dolphins, and sea lions, among other animals. Several habitats are refuges for threatened or endangered species, or for injured or undersized individuals. The park plays a very active role in the preservation of green sea turtles, Humbolt penguins, and Hawaiian monk seals.

Employment, Intern, and Volunteer Opportunities

If you already live in Hawaii, Sea Life Park may be your best bet for getting in the field. The staff is fairly large, consisting of about a dozen or so full-timers and a handful of part-time workers. The part-time staff are mostly students at the nearby University of Hawaii, which doesn't exactly constitute an internship program, but more of an unofficial relationship with the school. Full-time positions tend to open up once or twice a year. Although these positions have occasionally been filled by volunteers or part-timers, this doesn't necessarily happen often, and Sea Life Park is one of the few facilities where I would caution potential applicants from relying on extensive volunteer experience to provide a foothold. In fact, it isn't unusual for new trainees to be brought on with no experience in animal care at all. This is not to say that experience will work against you, but equal or greater emphasis seems to be placed on physical fitness and having a healthy attitude.

Although applicants needn't possess any remarkable skill with public speaking, they should demonstrate a willingness to learn. The positions which open up most often involve the park's theatrical presentation of local mythology, for which the park must hire locals. For this reason, coupled with the fact that all the Hawaiian facilities are leery of hiring mainlanders, local residents may hold a bit of an edge over other applicants. The only educational requirement for Sea Life Park is a high school diploma. Applicants should possess a SCUBA certification, or be prepared to obtain one.

An average work day at SLP is pretty typical for an animal park (or as close to typical as you can get surrounded by paradise), consisting of shows, training sessions, record keeping, and cleaning. The only noteworthy difference is Sea Life Park's use of "fish phantoms". No, not undead carp - the term describes the one staff member whose sole job is to creep into work at the ungodly hour of THREE in the morning and spend the next five hours bucketing wet, slimy fish all by himself. Then, as the rest of the staff arrives for the day, the Lone Fish Sorter takes a last appreciative glance at a job well done, and rides off into the sunrise, leaving behind a grateful but perplexed assortment of volunteers and co-workers to scratch their heads and ask, "Who was that smelly man?"

There are roughly two dozen dolphins and over three dozen sea lions at SLP. Although the staff is cross-trained to work with all the different animals, trainers tend to be assigned to a particular habitat, spending the majority of their days with its inhabitants.

Volunteers are occasionally used in animal sessions, but spend far more time cleaning. They are asked to commit to coming in one day per week for four months. After two consecutive four month periods, volunteers must sit out a cycle, to allow others to come in. Volunteering is a great way to get a feel for the job, and to get to know the staff, but may not be as helpful as a career path here as it is in many other facilities. Applicants for any Marine Mammal position should contact Marlee Breese, the Curator Of Mammals & Birds, or the Human Resources Department.

Research

SLP has supported a number of non-invasive research projects by outside researchers since the park's inception in 1964. The future of cetacean research at SLP is not completely planned out, and the potential exists for some projects to be accommodated. Qualified researchers may submit proposals to Marlee Breese.

Other Programs

SLP has a large number of classes and other marine education programs available, and each year nearly 40,000 people take advantage of them. Recent offerings include marine life discussions, animal training classes, nature hikes, overnight functions, and nature photography courses. For full details contact the Education Department.

Sea World

http://www.4adventure.com
(Email can be sent via the website)

Busch Gardens
P.O. Box 9158
Tampa, FL 33674-9158
Phone: (813) 987-5082
Fax: (813) 987-5443

Sea World of California
1720 South Shores Road
San Diego, CA 92109
Phone: (619) 222-6363
Fax: (619) 226-3964

Sea World of Florida
7007 Sea World Dr.
Orlando, FL 32821
Phone: (407) 351-3600
Fax: (407) 345-5397

Sea World of Ohio
1100 Sea World Drive
Aurora, OH 44202
Phone: (216) 562-8101
Fax: (216) 995-2117

Sea World of Texas
10500 Sea World Dr.
San Antonio, TX 78251
Phone: (210) 523-3600
Fax: (210) 523-3299

Overview

Sea World Inc. operates five theme parks which display cetaceans: Sea World of Texas, Sea World of Ohio, Sea World of Florida, Sea World of California, and Busch Gardens in Tampa Bay, Florida. These make up five of the nine Anheuser-Busch Theme Parks, which are owned and operated by the Busch Entertainment Corporation in St. Louis, Missouri. With over 150 cetaceans and almost as many animal care positions, Busch Entertainment by far constitutes the largest cetacean display organization in the world. The theme parks are family oriented institutions with a heavy emphasis on entertainment. They have come to be synonymous in the public eye with dolphin shows in the United States, especially through the showcasing of Sea World's ubiquitous "Shamu", the killer whale.

Employment, Intern, and Volunteer Opportunities

Sea World is respected throughout the animal care and training industry as a leader in developing new and effective means of animal husbandry and training. Each park has an animal care staff and a training staff; an unusual approach within the industry. The parks are home to orcas, pseudorcas, Commerson's dolphins, bottlenose dolphins, Pacific white-sided dolphins, short-finned pilot whales, and beluga whales, as well as dozens of species of birds, mammals, fish, and reptiles.

Sea World and Busch Gardens do not offer volunteer opportunities in animal care, but internships for college students are occasionally available. Contact the nearest facility for more information.

Research

Sea World has conducted numerous research programs over the past three decades, and contributes much to our scientific knowledge of cetacean physiology, behavior, and husbandry. Research programs are often being developed and introduced, often through the company's affiliate organization, Hubbs Research Institute. (See page 181.)

Other Programs

Sea World has made concerted efforts to contribute to conservation efforts across the country. It offers a number of educational programs within its parks, inside schools, and even broadcast them via satellite. Day-classes are available on a wide variety of marine subjects, and occasionally college-credit programs through local universities. Each park has its own Education Department, which provides listings of current offerings.

The four Sea World facilities are fairly active in the rehabilitation of stranded animals, including cetaceans. (Even the Ohio facility has had relatively stable animals brought in from the coast.) The threat of morbillivirus in recent years has reduced the park's ability to rehabilitate such animals, but sufficiently screened cetaceans are still taken in every so often.

Note bene: Sea World, Inc. declined to participate in this project on the grounds that the information I desired to publish was "proprietary". (This should give you an idea of the corporate climate which is so prevalent within the organization.) Consequently, this description is less informative than it should be, and for this I apologize. Hopefully, a more complete description will find its way into future editions. Until then, your best bet for up-to-date information about Sea World is to call them directly, or visit their website.

John G. Shedd Aquarium

1200 S. Lake Shore Dr.
Chicago, IL 60605
Phone: (312) 939-2426
Fax: (312) 939-2216
http://www.sheddnet.org

In a town full of internationally renowned architecture, world-class universities, rich heritage, and breathtaking skylines, Chicago's John G. Shedd Aquarium stands out as one of the city's proudest achievements. The Aquarium successfully blends education with entertainment and classical architecture with modern innovation, resulting in a marinelife exhibition and learning center second to none.

Overview

The John G. Shedd Aquarium is the largest indoor aquarium in the world. It houses close to eight thousand fish, birds, invertebrates, reptiles, and mammals. The original aquarium building was funded by John Graves Shedd, a philanthropist involved with dozens of city planning groups and charities. Sadly, John Shedd died before construction was underway, but through the efforts of his family members and the Shedd Aquarium Society, his vision was realized. The enormous Greek-inspired structure is festooned with aquatic adornments, and is part of Chicago's rich architectural history.

In 1991, the Shedd Aquarium once again set new records for style and sheer size when it unveiled its $42 million addition, the Oceanarium. Designed to embrace the concept of zoos and aquariums as vehicles for environmental protection and education, the new addition enclosed a small piece of the Pacific Northwest coastline, recreated within the Oceanarium's three-story walls. Winding paths lead visitors through various coastal habitats, offering a glimpse of the whales, dolphins, seals, otters, and...well, penguins of the Pacific Northwest. Although the Oceanarium was designed to showcase one specific biome, one of the project's major contributors had a thing for penguins. (They ARE pretty cute.)

The Oceanarium's three million-gallon closed system supports four Alaskan sea otters, five beluga whales, four harbor seals, six Pacific white-sided dolphins, and just under forty penguins. The building also houses two restaurants, an amphitheater, and an auditorium which usually offers one or more films on a variety of marine subjects. The marine mammal presentations have an educational emphasis, and are supplemented by numerous informative narrations throughout the day beside the otter and penguin habitats.

Employment, Intern, and Volunteer Opportunities

The Shedd employs more than three hundred people in its many departments, and its marine mammal staff is one of the largest in the world. Nearly twenty full-time animal care specialists work with the aquarium's whales, dolphins, seals, penguins, and otters. Animal care staff members are responsible for habitat maintenance, public narrations, and stocking the department's gigantic freezer in addition to the feeding, health maintenance, and training of the animals. The staff spends a great amount of each day in wetsuits, often feeling very cold from immersion in the habitats' 55° water.

The Shedd is very selective when hiring for full-time positions, even when compared to other display facilities. While entry-level positions are not uncommon, applicants are expected to possess a college degree in the life sciences or psychology, a SCUBA certification, and animal care experience of some kind. In addition to these tangible criteria, the Shedd searches for individuals who can demonstrate maturity, responsibility, good teamwork, and an understanding of the role that such aquaria play in education and conservation. The Shedd may make allowances for exceptional candidates which lack one or more of these criteria, but a strong applicant will possess them all. Interestingly enough, though the Shedd is one of the pickiest institutions, it hires more animal care specialists each year than most other facilities, usually two or three. Many positions have been filled by prior volunteers and interns.

With such a large collection of animals, the Shedd Aquarium often experiences periods of increased activity, such as cetacean pregnancies, new arrivals, and heavy penguin hatching seasons. The Aquarium will occasionally supplement its full-time staff with temporary workers in such instances. It is nearly impossible to predict when such positions will open, and they are usually given to prior interns or volunteers. Nevertheless, for the right applicant who happens to be in the right place in the right time, such a position could provide an invaluable opportunity for hands-on experience.

The Aquarium's volunteer force far outnumbers its paid staff, and the Marine Mammal Department has nearly eighty volunteers assigned to it. Twenty-seven of them are SCUBA diving volunteers who report at seven in the morning and prepare for a chilly hour-long dive in the Oceanarium's main habitats. They and the staff diver to which they are assigned usually spend that time cleaning algae, debris, and fecal matter from the habitat floors. Hoo-hah. The whales and dolphins are almost always moved into an adjacent habitat for this procedure, though the divers usually get a close look at the animals through the mesh gates. Applicants for volunteer diver positions are required to have at least some cold water diving experience, and preferably a rescue-diver certification as well.

The department's fifty-two other volunteers assist with food preparation, cleaning, habitat maintenance, and occasionally help the trainers during animal sessions. Volunteers are not guaranteed interaction with the animals, but are often taken on closely supervised play sessions, which can sometimes include limited animal contact. Each day a morning volunteer shift begins at seven, followed by an afternoon shift at about one o'clock. Each shift lasts about six hours, and consists of two volunteer partners who work together from week to week. Volunteers are asked to commit to one year's worth of service in a consistent weekly shift. Weekend volunteers come in once every other week. There is a waiting list for all volunteer positions in the Marine Mammal Department, and many people wind up volunteering in other departments until a position opens up.

Internships in the Marine Mammal Department last about three months, and are intended for college students with an interest in animal care, training, veterinary science, or a related field. Non-students may apply, but will be at a slight disadvantage during the selection process. The internships are unpaid, entail 25-30 hours of work each week, and provide participants with one of the industry's best exposures to the field of marine animal care and training. Prior experience in animal care is not necessary, though the Shedd looks for the same characteristics in potential interns as it does in applicants for paid positions.

The Shedd offers twenty-four internships each year and are divided evenly into winter, spring, summer, and fall seasons. Applications for all seasons in a given year are taken no earlier than October 1st of the preceding year, though establishing and maintaining contact with the department should be done as soon as possible.

Research

The Shedd Aquarium has often participated in marine mammal research conducted by outside investigators, including hearing threshold projects, behavioral studies, and the development of EKG equipment and tracking devices for use with dolphins and whales. While the Aquarium does not initiate many studies of its own concerning cetaceans, it would be happy to consider proposals from qualified researchers.

Other Programs

The Shedd offers a host of educational programs both for members and to the general public. One-day classes are available year-round on just about any marine topic imaginable, including a number of programs focusing on marine mammals. Through a partnership with Western Illinois University, the Aquarium often offers undergraduate and graduate level courses on general marine mammology, bioacoustics, and animal care and training. Outreach programs, teaching curricula, discounted admissions, and group activities are available for school groups. The Shedd also offers a wide range of field trips, both locally and around the world, which provide opportunities to learn, travel, and enjoy yourself all at the same time. For information on any of these programs, contact the Education office.

Theater of the Sea

PO Box 407
Islamorada, FL 33036
Phone: (305) 664-2431
Fax: (305) 664-8162

Overview

Originally laid out near the turn of the century as a portion of the disastrous Henry M. Flagler railroad project, Theater Of the Sea is the second oldest marine park on the planet. A haven for sick and injured animals, the facility is also home to bottlenose dolphins, sea lions, sharks, and a host of other indigenous marinelife.

TOTS was established in 1946 by the P.F. McKenney family, who still own and operate the park. The Theater's dolphins live in three acres of natural lagoons, which are fed by twelve million gallons of ocean water daily. TOTS operates one of America's four dolphin "swim-with" programs, and has made it available to local trauma patients, as well as to the "Make A Wish" program for terminally ill patients.

One of the absolutely coolest things about TOTS is its massively huge number of cats. They're everywhere. Long ago, the Theater was naive enough to take in a cat being left behind by a local resident. Like many people who move into the keys prematurely, he'd had enough of island life and was headed north. Pretty soon, word spread through the Keys' IMOUTTAHERE network, and carload after carload of mainland-bound families started dumping off their cats. There are now about a hundred abandoned cats throughout the facility, which makes the place worth visiting all by itself.

Employment, Intern, and Volunteer Opportunities

TOTS's fourteen animal care staff specialists are divided into four teams. The Land Show team handles the park's fish and birds, the Sea Lion Show team handles the sea lions, the Dolphin Swim group conducts the dolphin swim program, and the Dolphin Show team, well, you can probably guess what they do. (Everyone works with the cats.) Positions open up about once per year on average, though the frequency fluctuates from year to year. TOTS seems to hire new employees from the outside world as well as from its own volunteer and intern ranks, though almost all employees arrived with animal care experience of some kind. A college degree is not mandatory, though it is recommended. SCUBA certification is desirable.

The internship program is designed for college students looking for two to four weeks of practical experience. All interns gain experience working in the dolphin and sea lion shows. Three-week interns also work in the land show area, and four-week interns get to help with the swim program. The program requires participants to write up a paper on their experiences in each department. Interns learn animal care and training, get hands-on experience with the animals, and get to swim with the dolphins at least once during their stay. TOTS participates in a reciprocal internship program with the Dolphin Research Center and the Dolphin Connection, sending their interns to each facility for a day.

TOTS has no formal volunteer program in place, though the staff might be able to find particularly persistent locals something to do. (Opening cans of cat food, I would think.)

Other Programs

People wishing to swim with the dolphins at TOTS should call ahead as far as possible, and will be required to give 50% of the $85 fee as a deposit, which is refundable up to seven days before the scheduled date. The price includes a thirty-minute briefing, a thirty-minute swim, and admission to a number of animal shows.

TOTS's "Trainer For a Day" program lets people over 10 experience the excitement, the prestige, and the nasty, fishy stench that is modern marine mammal care. Participants are taken on animal sessions, shown how to prepare food, and are given a briefing on the essentials of animal care and training. The entire affair lasts about five hours (including shows), and is a bargain at $75.00 for young adults thinking about getting into the business.

The Theater offers the "Dolphin Adventure Snorkel Cruise", a four hour tour of nearby islands and reefs.

A four hour tour...

A four hour tour...

Sorry. Participants learn about the history and ecology of the area, take in the sights, and have the option of snorkeling through the reefs. Large groups are welcome, and snorkel gear is provided. The trip costs $49.95 for adults, and $29.95 for children 2-12. (Kids under 2 go free.)

United States Navy

Space and Naval Warfare Systems Center San Diego
(SPAWAR Systems Center San Diego)
San Diego, CA 92152-5001
Phone: (619) 553-2724
http://www.nosc.mil/nrad/welcome.page
Tom LaPuzza - Public Affairs Officer
lapuzza@nosc.mil

Ah, the Navy. It's always been easy, even for relatively well-informed people, to imagine the military as the Bad Guys, strapping bombs to dolphins' backs and using sea lions to root out mines. There will always be a few conspiracy nuts out there who will simply believe whatever happens to fit in the uniquely shaped void between their ears, but the rest of you, I hope, will believe me when I say that the Navy does not use any animals as any kind of weapon whatsoever, nor does it have any plans to do so. Because of the interest surrounding this topic, three separate researchers approached it from different angles, pursuing the Truth at many separate echelons. I might also add that we're all X-Philes here at Omega, so you can believe we jazzed up in Hyper-Motivation Mode. We interviewed Navy PR, the training and research staff at San Diego, enlisted personnel involved in the end products of the research, prior servicemen with nothing to lose, and an ex-Navy dolphin, who wouldn't talk, now retired in the Florida Keys. Everyone came to the same conclusion: the animals in the Navy program are used strictly in non-offensive roles. Now that we've got that out of the way, let's get to the details.

Overview

The Navy has just over a hundred marine mammals, including dolphins, sea lions, three beluga whales and one false killer whale. Most of these are in San Diego, CA, though a handful are in other locations, which we'll get to in a bit. There are three components of the Navy that work with these animals. These are: Navy civilian scientists and veterinarians who manage the program, perform the research and development, and direct animal care; uniformed Navy personnel who run the operational systems; and a private company which provides training and animal care services under Navy contract. The current contractor is Science Applications International Corporation (SAIC). All these are supported by

Navy/Marine Corps military and civilian executives in Washington who make the high-level decisions about which marine mammal capabilities will meet Navy requirements and who will provide funding to pay for the research, the operational systems, and the care and feeding of the animals.

The Navy civilian scientists and program managers research potentially useful marine mammal capabilities. After initial animal training and development of associated required hardware by Navy civilians and contractors, marine mammal systems are then turned over to uniformed Navy personnel for day-to-day continued training and operation. Typically, once a system is handed over to the operational Navy, one experienced Navy civilian or contractor will work with the operational unit to supervise and coordinate the Navy military handlers. This entire program is run by one of the Navy's major research, development and engineering centers.

I wish I could say which center, but its name changes on a weekly basis. I've put a detailed history of the San Diego center's name at the end of this entry, in case any acronym freaks are reading. The rest of you can probably call the place SPAWAR for the near future, though I'm sure it will mutate again before long.

The Navy's Marine Mammal Program started nearly forty years ago as an effort to analyze the animals' hydrodynamic capabilities, and to determine if the same principles could be applied to naval ships, submarines, and torpedoes to increase their speed using the same engines. Marine mammal physiologists of the period believed these animals did not have the muscle structure and power for their observed speed, and theorized as one possibility their possession of some means of reducing drag that would otherwise reduce their speed to predicted levels.

Those early studies yielded mixed results, and little or no data led to the improvement of vessels' movement through water. It did become apparent that a great deal of helpful information lay in the animals' use of biological sonar. The investigators were also impressed with how trainable and dependable the animals were. The Navy's research split into two directions. One path investigated whether or not the animals could perform basic tasks under adverse conditions, such as in very deep, dark, or cold water. They also set out to determine whether or not such tasks could be reliably performed in the open ocean. The second path tried to artificially reproduce cetacean sensory systems.

The computers and neural networks necessary to reproduce biological sonar just aren't available (yet), but the research and operational systems development have met with tremendous success. In addition to Navy pioneering research into all aspects of marine mammal science, Navy cetaceans can locate, mark or attach recovery hardware to objects on the sea floor. Navy dolphins and sea lions routinely perform object search and recovery at 500-1,000-foot depths, at least several times deeper than is customary for human divers. A previous experimental effort involving pilot whales demonstrated the feasibility of recoveries to 1,654 feet. Dolphins and sea lions can carry cameras underwater to show surface personnel what they've found. Dolphins can even patrol a perimeter around a pier or a ship at anchor and alert their handlers if an unauthorized swimmer is in the area. In both research and training, much of the Navy's work is ahead of its civilian counterparts'.

Although the Point Loma area of San Diego houses most of the Navy's marine mammals, a few are stationed on "active duty" in Charleston, SC, and a handful more are on loan to the University of California, Santa Cruz, and the University of Hawaii facility on Coconut Island, Hawaii.

The Navy relies on captive-bred animals to replenish its population as older animals are retired. Retired animals continue to be well cared for, and live out the rest of their days in San Diego, much like many humans. This enables the Navy to join the increasing number of facilities with voluntary moratoriums on wild captures of marine mammals. In fact, by allowing qualified facilities with capture permits to take on surplus animals (no, you won't find them in Army/Navy surplus stores) the Navy has alleviated the need for many other groups to turn to wild captures.

Volunteer, Intern and Employment Opportunities

It's almost unnecessary to say that there are no volunteer opportunities in the Navy's program. Of the three groups dealing directly with the Navy's animals, the Navy civilian scientist group is the most difficult to get into. Positions fall under U.S. Government Civil Service jurisdiction, and there are currently only about 26 scientists (biologists, psychologists, and physiologists), veterinarians and technicians employed by the program. Nearly all of them have 20-25 years of experience conducting research and development projects with marine mammals. When new civil service employees are hired, which happens occasionally, they typically need to have many years of applicable experience to qualify.

Becoming a marine mammal handler in the uniformed Navy is

very difficult, and I don't recommend trying to do so if you aren't already an enlisted Navy diver. No matter which branch of the Service you join, there are jobs you can enlist for and have guaranteed in your contract, and then there are "specialty" jobs which are more of a luck-of-the-draw kind of thing. Animal handling is an example of the latter. There are only four or five active "systems" of animals at once, each of which is assigned to an Explosive Ordnance Disposal (EOD) unit. EOD units, in turn, fall under one of several diving commands in the Navy, which is probably the most specific size of unit you can be sure of getting when you enlist as a diver. So IF you can sign a contract with the Navy for diver school, (and graduate) and IF you apply for and get to attend the Navy's one year Explosive Ordnance Disposal school, and IF, after all that, you happen to be assigned to an EOD unit with a marine mammal system, then maybe, MAYBE, you'll be able to bribe someone to assign you to it. So do yourself a favor, don't walk into a Navy recruiting office and say "I want to train dolphins". You'll wind up a being a cook on a cruiser somewhere north of Labrador.

If you feel you absolutely HAVE to get involved with the Navy's marine mammal program, their civilian contractor is your best bet. See page 123 for information on SAIC.

Research

They don't always get the credit for it, but the Navy is responsible for much of the early research done on dolphins and whales, and most studies done today in the areas of cetacean acoustics or physiology depend heavily on Navy groundwork. Navy civilian scientists continue to contribute heavily to the scientific literature on marine mammals. (The Navy's bibliography of publications on this program is currently under revision, to add the more than 100 new papers published in the last five years.) The very first Navy study of a cetacean took place in 1960, and was done with a Pacific white-sided dolphin named Notty, who failed (for better or worse) to teach the Navy how to build a faster torpedo. Since then naval research has included studies of marine mammal physiology and acoustics, techniques for diagnosis and treatment of health problems, non-invasive neurophysiological studies into the functions of the cetacean brain, and the development of advanced animal care and training techniques. The Navy's marine mammal scientists publish scientific papers in the areas of marine mammal communication, health care, breeding, behavior, hydrodynamics, and open sea release.

A History Of The San Diego Research Facility

Skip this if you don't find names terribly interesting. The San Diego center, with many of the same personnel, buildings, and mission to perform for the Navy, has changed its name six times during the nearly four decades it has managed the Navy's Marine Mammal Program. In 1959 when the program began, it was the Naval Ordnance Test Station (NOTS). Reorganization in 1967 led to the name Naval Undersea Warfare Center (NUWC), but with the unpopularity of the term "warfare" in the mid-Vietnam War period it became the Naval Undersea Research and Development Center (NURDC) in 1969, and, to simplify, the Naval Undersea Center (NUC) in 1972. Five years later, NUC consolidated with its neighbor Navy laboratory on Pt. Loma in San Diego, the Naval Electronics Laboratory Center (NELC), to form the Naval Ocean Systems Center (NOSC). In terms of longevity and its relatively unique and wide-spread e-mail addresses (nosc.mil), this was the most popular name, lasting until 1992.

That year, the Navy reorganized its seven major research and development centers, including NOSC, and 29 engineering centers into four warfare centers. NOSC was the major component of the warfare center called the Naval Command, Control and Ocean Surveillance Center (NCCOSC), and was re-named the NCCOSC Research, Development, Test and Evaluation Division (NRaD), a name which stuck until a few short weeks ago. As a result of the Base Closure and Realignment Commission (BRAC) process back in 1995, NCCOSC was disestablished on September 30 of this year, and the NRaD organization became the Space and Naval Warfare Systems Center San Diego, SPAWAR Systems Center San Diego for short. At the moment this is being written, NRaD is still on all the buildings, but look for it to come down shortly in favor of something probably like SSC SD.

You pay these people with YOUR taxes, guys.

Vancouver Aquarium

P.O. Box 3232
Vancouver, BC V6B 3X8
CANADA
Phone: (604) 685-3364
Fax: (604) 631-2529

Way back in 1888, Vancouver's city fathers recognized the value of public park lands and sectioned off a huge peninsula jutting out into English Bay. They designated it as a public park, and took steps to protect it from future development. The 405,000 acre Stanley Park, as it was named, remains one of the most beautiful urban parks in the world. Residents and tourists alike flock in droves* to visit the park's dense forests, monuments, bike paths, seawalls, and beaches. The park also features turn-of-the-century lighthouse, an outdoor theater, and Canada's Pacific National Aquarium, the Vancouver Aquarium.

Overview

The Vancouver Aquarium is Canada's first and largest aquarium, and continues to be internationally recognized as a leader in animal care, education, and research. Among its many species, the aquarium houses harbor seals (oh, excuse me, harbour seals, since they're in Canada) Steller sea lions, sea otters, killer whales, and one Pacific whitesided dolphin which thinks it's a killer whale. (It grew up around them.) The aquarium gives educational orca presentations, and uses some of the best displays and exhibitry in North America.

Employment, Intern, and Volunteer Opportunities

As with all Canadian facilities, Americans won't have much chance of getting started in a paid position here. For Canadian applicants, the Marine Mammal Dept. prefers a college degree, preferably in the biological sciences. A driver's license is necessary, and a SCUBA certification is a good idea. Most staff members come from the Aquariums's volunteer ranks, or have volunteer/intern

*If you've never flocked in a drove before, I highly recommend it. It's good for the circulation, but make sure to drink plenty of water.

experience at another zoo or aquarium. There are about a dozen animal care staff, and one or two vacancies open up every year or so.

Americans may apply for internships at the aquarium, though there is no formal internship program. Your best bet is to learn as much as you can about the facility and make a proposal for a project which would fit in well with the department's schedules and policies. Internships in the past have been unpaid.

The Marine Mammal department uses volunteers, though not in direct contact with the animals. Volunteers clean, prepare food, maintain records, etc. Volunteers assist in rehab work for stranded marine mammals, which sometimes involves hands-on experience. Rescue and rehab work is open to volunteers who have been with the aquarium for a minimum of one year. Volunteer applicants may be from the United States, but as they require a one year commitment (one morning per week), you would have to have to reside reasonably close to Vancouver.

Research

The Vancouver Aquarium works hard through in-house research to increase the scientific community's understanding of dolphins and whales. Studies include cetacean reproductive cycles, diet and energetics, hearing thresholds, and and communication techniques.

The Vancouver Aquarium is one of three institutions conducting extensive studies on the Pacific Northwest's resident and transitory orcas (the other two being Orcalab in Alert Bay, British Columbia, and the Center for Whale Research in Friday Harbor, WA.) This work is being headed up by Vancouver Aquarium's Dr. John Ford, a renowned marine mammal scientist and an expert on the orcas of the Pacific Northwest coast. Although volunteers are not really used to support this program, Dr. Ford does occasionally recruit graduate students from the nearby University of British Columbia (see separate entry.)

Other Programs

The Vancouver Aquarium offers a huge selection of one-day classes, seminars, curriculum-based school programs, excursions and field trips, and other educational experiences. For a full listing, contact the aquarium's educational office. Memberships and orca adoptions are also available, as well as a host of informational leaflets on marine biology and related careers.

West Edmonton Mall Dolphin Lagoon

2872 8770 170th St.
Edmonton, Alberta T5T 3J7
CANADA
Phone: (403) 444-5346
Fax: (403) 444-5266

West Edmonton Mall. Gather, ye shoppers, and bask in that which is the glory of All Things Purchased! West Edmonton Mall! Vast tracts of thick, gooey consumerism dripping through endless corridors of hypercommerce, giving way to the blinding, all-penetrating glare of Pure Barter! The walls reverberate with the Beanie-Babyish power of Product which holds us in its grinning, mesmerizing sway. You know you want it. You've gotta have it. Lord knows what it is, but you can get it here! This is IT! The Big One! West...EDMONTON...MALL!!!!

(whew.)

Overview

Okay, I don't mean to be quite so sarcastic, (well...yes I do) but this thing just has to be seen to be believed. Even larger than the Mall of America in Minneapolis, West Edmonton Mall has over 800 stores and services, 11 major department stores, the Palace Casino, an amusement park, a giant wave pool, Ceasar's Bingo Hall, 110 eating establishments, 19 movie theatres, a Deep Sea Adventure attraction featuring four real submarines, several coral reefs featuring more exotic fish than you can shake a stick at (I tried) and, of course, a Dolphin Lagoon. As the world's largest shopping and entertainment complex, it is a temple to capitalism unrivaled in the known universe - the Mecca of Malls.

Four Atlantic bottlenose dolphins reside in the mall's Dolphin Lagoon. The Lagoon is one of three divisions of the Marine Life Department which includes the mall's aquariums and penguins. When visitors get tired of shopping, they can view over 200 species of fish, reptiles, and exotic birds.

The Dolphin Lagoon's staff does a surprisingly good job of working educational content into their presentations. Constantly changing narrations touch on dolphin anatomy, physiology, natural history, and conservation issues. While some may question the ethics of staging a dolphin habitat inside a shopping mall, it would seem the mall's program has been responsibly executed and well maintained.

Employment, Intern, and Volunteer Opportunities

The Dolphin Lagoon uses about fifteen volunteers to assist with cleaning, food prep, and public interpretation. Volunteers must be over eighteen years old, and are often undergraduate students studying within the biological sciences at the nearby University of Alberta, though anyone may apply. Volunteers are requested to give one four-hour day every two weeks with no minimum duration. Because so many volunteers tend to be students, it is always easier to get a position in the summer.

Interns are not used at the lagoon just yet, but they are considering implementing an program. There are eight full-time trainers on staff, some of whom work with more than one animal group. Openings appear once every few or so, on average. The lagoon employs four public interpreters, which could be a nice way to get your foot in the door. As with all Canadian facilities, paid opportunities for Americans are almost nonexistent.

Research

The Dolphin Lagoon has no ongoing research programs. In the past, it participated in simple physiological studies. They would consider proposals for projects which would not place undue stress on the animals or staff.

Other Programs

The Mall offers a several educational classes for pre-school through twelfth-grade students on weekday mornings and afternoons. For listings call the Group Sales office at (403) 444-5386.

5
Research Organizations

It's even harder to cover cetacean research jobs than dolphin and whale care and training. One animal care job looks fairly similar to another, but research positions? Oy. No two are the same. They're scarce, scattered throughout numerous fields, poorly funded, and not always stable. You can't approach cetacean research as you would becoming an accountant, or lawyer. Let's take a look at what kinds of jobs exist. Then we can worry about how to get them.

What is cetacean research?

There are many different kinds of cetacean research projects, but most of them fit into one of two general categories. The first deals with large groups of animals in the wild, something many people refer to as conservation biology. By far the most common, this type of cetacean research involves mapping out where dolphins and whales live, how many there are, and where they travel to. Conservation researchers need to know how animal populations are affected by pollution, habitat encroachment or destruction, fishing, and boat traffic. Photo-identification and quantitative analysis, the two most powerful tools in conservation studies, enable researchers to identify troubled populations of dolphins and whales.

Researchers using photo-identification can keep track of wild animals without using invasive techniques like radio tags, identification markers, or freeze-branding numbers into an animal's skin. Scientists compare photographs to catalogs of known individuals, and use distinguishing colorations, marks, or disfigurements to tell which animals were seen.

Once they figure out where these animals are, they use quantitative analysis to draw a picture of the population as a whole. By

(Photos courtesy of Keith Rittmaster, North Carolina Maritime Museum)

Photo-identification studies use distinguishing characteristics, like the notches on these dorsal fins, to keep track of individual animals and entire populations

tracking populations over time, researchers can draw inferences about how the animals are affected by man. It's an excellent system, but it also means many researchers spend long hours in darkrooms, comparing photographs, entering data, and carrying out complex mathematical operations. Even data collection can test a researcher's stamina, both in cold and rainy regions as well as out in the hot, burning sun.

Not all cetacean research involves population studies. I'll refer to the other general category as "laboratory research". By "laboratory" I might mean the sort of test tube and Bunsen burner conglomeration you'd automatically think of, or I may have an isolated habitat at a display facility in mind. I could even be talking about a stretch of the open ocean, but in all three cases, the research focus-

(Photo by Laurel Canty Ehrlich and provided by DRC)

Research may be carried out on an entire population, or on individual animals

es on a particular aspect of cetacean biology. How many species of bottlenose dolphins are there? How do whales swim so fast? How high or low does cetacean hearing go? What purpose do humpback whale songs serve? How smart are dolphins? Although quite a few organizations and individual scientists try to figure out stuff like this, figuring out where those people are requires a little homework.

Who conducts cetacean research?

This chapter shows you many facilities which use cetacean researchers. Groups like Allied Whale, the Cetacean Behavior Lab, Moss Landing Marine Laboratories, and the Marine Mammal Research Program at Texas A&M combine graduate education with actual field research. Ocean Mammal Institute, Earthwatch, Coastal Ecosystems Research Foundation, and Oceanic Society Expeditions specialize in giving the general public an opportunity to participate in research programs for a fee. ORCALAB, the Cetacean Research Unit, and the Center for Coastal Studies manage to support themselves independently while they carry on their research. Some studies are the pet projects of individual researchers, like Project Pod, or the Nag's Head Dolphin Survey.

Many organizations in other sections of the book carry on cetacean research projects. Research is the main function of the Dolphin Institute and the Hawaii Institute of Marine Biology from chapter four. Lots of groups from the conservation section conduct or fund research as well, including Cetacean Society International, Earthtrust, and the International Wildlife Coalition.

The government commissions a number of marine mammal studies each year, some of which focus on cetaceans. The National Marine Fisheries Service, the Marine Mammal Commission, the Navy, and the Minerals Management Service do so more often than anyone else.

Cetacean researchers with a little experience might receive employment offers from colleges, museums, zoos, aquariums, environmental groups, or even oil companies. Many opportunities present themselves to scientists which establish themselves within the field.

How can I become a cetacean researcher?

First thing's first. Before you can dash off and start counting whales for the government or the like, you need to become a capable scientist. Focus on building an undergraduate education in gen-

ence while getting some kind of experience with the kinds als you like. Don't worry so much about finding an under- e program which offers marine mammal courses. A few exist, and if you manage to get in, fine. Otherwise, just find a college you like which is near a coastline or a display facility with a research program.

Pick a major like biology, zoology, chemistry, physics, or psychology. Take some courses in quantitative analysis, statistics, and computer science. Chances are your school will offer at least one photography course. Take it too. Scientists do a lot of writing, so do as well as you can in your English classes, and look for a scientific writing course.

Your spare time is one of your most valuable tools. You won't learn much about dolphins and whales from your undergraduate classes, so check out the book list on page 353 and start reading. Become a student member in the Society for Marine Mammology, and start reading their journal, Marine Mammal Science. Marine mammal research is published in other journals as well. If you keep your eye on the Society's journal and Biological Abstracts (which you'll find at most major libraries) you'll have an idea for what kinds of projects are being done and who's carrying them out.

Getting some first-hand experience is just as important. Look through the organizations in this book and see if you can't volunteer your services for a research program somewhere. Take trips during your summer breaks and see what kinds of studies, species, and locations are out there. If you're active enough during your first four years of college, you'll have no problem starting the next step: grad school.

Grad school is where you'll have your first chance to apply your new scientific skills to cetacean issues. There are a number of schools with graduate-level marine mammal programs. (See page 350 for a list of them.) If you can't get into one of these, you'll have to find a graduate advisor with an interest in marine mammals. Check those journals, read the books, and go through the facilities in this book. Go to conferences held by the ASA, SMM, IMATA, and ACS. (See Chapter Six.) You'll have a good working knowledge of which cetacean researchers share your interests, and which universities they're affiliated with. The Society for Marine Mammology's pamphlet Strategies For Pursuing a Career in Marine Mammal Science offers some good advice about finding an advisor:

> "Students should realize that...graduate school faculty do NOT have to advise students just because they are enrolled at their university. Students sometimes enroll at a university because of a well-known professor and assume they will have the opportunity to work under him or her. BEFORE entering a graduate program, contact the professor and establish his or her willingness to serve as an advisor."

Introduce yourself to the person you'd like to have as an advisor, whether in person or by mail, and tell them what your interests are. Ask them about their work, the school they're affiliated with, and whether they would be willing to be your advisor. Some of the college-affiliated researchers in this book use volunteers in their work. Getting involved in a project before grad school can make it much easier to obtain sponsorship.

Cetacean science and research is a diverse, unpredictable, and ever-changing field. As with the display industry, no two scientists can tell the same story. There are no rules in this business, and ingenuity, persistence, and hard work are your most valuable tools.

Research Organization Descriptions

On the following pages are descriptions of twenty-seven organizations in the U.S. and Canada whose primary reason for being is cetacean research. As mentioned above, Chapter Four lists two more such organizations: HIMB and the Dolphin Institute. Each description is divided into four sections. "Overview" gives a general background and description of the group. "Research Projects" tells you what the heck they're up to. "Involvement Opportunities" replaces "Employment, Intern, and Volunteer Opportunities" from Chapter Four. Your main goal at this point should be to gain experience, not employment. "Other Programs" describes helpful or interesting activities each group offers in addition to its research.

One last note before you dash off: research programs change far more often than display facilities. Most of the groups described in this book are pretty stable, but don't assume there aren't other projects out there. Keep reading, and get a membership with a couple of the organizations in Chapter Six. The newsletters will help you keep abreast of current cetacean issues. If you can, buy copies of the IUCN Cetacean Action Plan (page 240) and the Marine Mammal Commission's Annual Report (page 260). The more you know about current cetacean environmental and political issues, the better idea you'll have of where research is likely to be funded.

The Alaska-British Columbia Whale Foundation

c/o Behavioural Ecology Research Group
Department of Biological Sciences
Simon Fraser University
Burnaby, B.C. V5A 1S6
CANADA
Phone: (604) 291-4374
Fax: (604) 291-3496
http://mendel.mbb.sfu.ca/berg/whale/abcwhale.html

Overview & Research Projects

The Alaska-British Columbia Whale Foundation is an excellent example of how a highly focused machine can excel at accomplishing a single task. Under the direction of its two primary researchers, Fred Sharpe and Larry Dill, the Foundation's primary purpose has been the investigation of the highly complex cooperative feeding strategies of humpback whales. Using sonar imaging, genetic and acoustic analysis, tagging, laboratory experiments, and computer simulations, the Foundation is trying to unravel the mysteries surrounding the whales' use of bubble nets, loud trumpeting noises, and communication between one another in the rounding up and snarfing down of entire schools of fish. (For more on humpback cooperative feeding, see the discussion of cetacean intelligence on page 19.)

Involvement Opportunities

The Foundation makes use of a large number of volunteers in its research efforts. In the summer, volunteers may accompany the research staff to Alaska, where for 2-4 weeks they assist in data collection and recording, photography, and data entry. During other seasons the volunteers are needed to help in photograph matching, data analysis, etc. Volunteers who assist with data analysis and vessel maintenance during the non-field season (October to May) receive priority for assisting with the at-sea studies (June through September). While there are no set criteria for potential volunteers, experience in quantitative analysis and multimedia applications is very helpful.

Allied Whale

c/o College of the Atlantic
105 Eden Street
Bar Harbor, ME 04609
Phone: (207) 288-5644
Fax: (207) 288-4126
alliedwhale@ecology.coa.edu

Overview

The Allied Whale group at the College of the Atlantic was formed in 1972 by a group of faculty, students, and staff members. Its purpose was to provide a platform for research on the whales of the North Atlantic. Allied Whale maintains the photo identification catalogs for both the North Atlantic humpback whale and the North Atlantic finback, or fin, whale. In constructing these catalogs, Allied Whale pioneered a number of photo identification techniques which are now being used by cetacean researchers worldwide. Allied Whale scientists have been involved in many collaborative projects in other parts of the world, including Canada, the Antarctic, New Zealand, the Dominican Republic, and Bermuda.

Research Projects

The majority of Allied Whale's research focuses on the distribution patterns and population dynamics of whales in the North Atlantic ocean. This is accomplished through boat-based observations and photography, followed by photographic analysis and cataloging on the mainland. Allied Whale is in the process of digitizing its enormous catalogues, which contain pictures of over seven thousand individual whales. In addition to the humpback and finback catalogues, Allied Whale has begun yet another Herculean task: the cataloguing of humpbacks in Antarctic waters.

Involvement Opportunities

While Allied Whale cannot accommodate volunteers or interns from outside the College of the Atlantic, there are quite a few opportunities for interested students at COA. During the school year, just under a dozen interested students from various departments are offered the chance to help with photograph identification and quantitative analysis of the whales' populations. During the summer, when most of Allied Whale's field research is carried out, several students are usually taken along to assist with data collection aboard research vessels. Most students assisting with Allied whale's programs obtain college credit for their efforts, or else use the experience to qualify for the financial aid department's work/study program.

Other Programs

Allied Whale provides a number of educational resources to the public, including informational leaflets, bibliographies, curriculum resource lists, slide presentations, lectures, and a biannual newsletter. Allied Whale's staff is very approachable, and pleased to provide any information it can to the public.

Atlantic Dolphin Research Cooperative

(no mailing address)
http://members.aol.com/adrcnet/index.html
Richard S. Mallon-Day - Website Supervisor
r.mallon1@genie.geis.com

The ADRC is a somewhat loosely knit group of Eastern Seaboard researchers who banded together to exchange information and ideas on a regular basis. The ADRC holds an annual conference, which can be a good place to meet researchers and listen to speakers outline the results of their efforts. Written transcripts of past presentations are posted on the ADRC website, which is not only educational, but can tell you a bit about what the various researchers are up to. Upcoming conferences and events as well as newsletters and other publications are also available there.

ATOC (Acoustic Thermometry of Ocean Climate)

University of California, San Diego
IGPP 0225
9500 Gilman Dr.
La Jolla, CA 92093-0225
Phone: (619) 534-8031
Fax: (619) 534-8076
Dr. Chris Clark - MMRP Hawaii
cwc2@cornell.edu
Dr. Dan Costa - MMRP California
costa@biology.ucsc.edu

Overview

For years now environmental scientists have been concerned about ozone depletion, industrial pollutants, and the possibility of "global warming". For the first time in recorded history, humans are affecting their environment on a global scale, and changes occur faster than we can document them. Despite dozens of studies by scientists from all over the planet, nobody can say for certain what is happening to Earth's average temperatures. The oceans have a direct relationship to the atmosphere, and accurate information from both is necessary to draw conclusions about global warming. ATOC is an attempt to provide some real climate data to study, so that scientists will be able to both test existing models, and use the same models or modified versions to make climate predictions. By testing and improving climate models now, ATOC can help make progress toward greenhouse predictions later.

The ATOC program is a feasibility study composed of two environmental initiatives which involve a number of scientists and institutions. The first goal of the program is to obtain ocean temperature data. This is done by measuring the time it takes for low frequency sound signals to travel from ATOC sources to a network of receivers. The second goal is to assess what effect these sound transmissions have on marine mammals, sea turtles and other marine life. The ATOC Marine Mammal Research Program, made up of a team of researchers from a number of organizations, provides information on the hearing capabilities of marine mammals and sea turtles, and the responses of marine mammals to ATOC sounds.

Most of ATOC's support has come from the Strategic Environmental Research and Development Program at the

University of California. The program is being coordinated by Scripps Institution of Oceanography. There are dozens of individual scientists and teams involved with the project in varying capacities all around the Pacific ocean.

Research Projects

It is the MMRP which dictates the transmission schedule of the ATOC sound source. There is currently one source in use, on Pioneer Seamount, an undersea structure 3,000 feet below the ocean's surface and about 50 miles offshore OF Half Moon Bay, California. Dr. Dan Costa of the University of California's Long Marine Lab (see related entry) is responsible for the MMRP in this area, and so leads the team which sets the California source's schedule. The Pioneer source has been active since December of '95.

In Hawaii, the MMRP is under the direction of Dr. Christopher W. Clark of Cornell University's Bioacoustic Research Program. Hawaii's program will begin in the fall 1997. During the winter of 1998, a major research effort will study the responses of humpback whales to ATOC sounds using the operational Kauai sound source.

Involvement Opportunities

There are many aspects of the MMRP which one could track down. Aerial and surface surveys are being made of large whales, and captive odontocetes and pinnipeds in some facilities are being tested for their hearing thresholds. The ATOC office may not have all the information on hand, as not all the research is funded by them, but if you can "network" (I hate that term) with researchers in the project you should be able to learn enough to get started. There has been some opportunity to participate in field research, however the project ends in fall 1998, so such opportunities are limited.

Cascadia Research

218 1/2 W. 4th Ave.
Olympia, WA 98501
Phone: (360) 943-7325
Fax: (360) 943-7026
cascadiaresearch@compuserve.com

Overview

Cascadia conducts quite a bit of cetacean research each year, often at the request of government departments and universities. Even now, Cascadia scientists observing from land and sea map out blue, fin, and humpback whale populations for the National Marine Fisheries Service. Twenty years of such projects form a nearly unmatched history of marine mammal research.

Cascadia materialized in 1979, the result of a little collaboration and initiative between several Evergreen State College graduates. The students had some research experience with NMFS's National Marine Mammal Laboratory, which put them to work right away. In the two decades since, Cascadia worked on contracts from other numerous state and federal offices, including the EPA, the Marine Mammal Commission, and the Army Corps of Engineers.

Research Projects

Any figures you may come across for North Pacific gray whales originate from Cascadia research efforts, and the group documents populations of blue, fin, and humpback whales. Cascadia investigates the impact of human pollution, traffic, and encroachment on marine mammals. It shares the responsibility of analyzing the effect of the ATOC sound sources on several marine species. Cascadia receives assignments on many cetacean issues, and merits a bit of scrutiny itself for enterprising marine mammal scientists.

Involvement Opportunities

Cascadia takes on interns to assist in its research. Interns come primarily from Evergreen State College, though others may apply. Participants spend much time printing and matching photographs. Once students build some experience with photo-identification, and if the season is right, opportunities to help with field work may arise.

Center For Coastal Studies

P.O. Box 1036
Provincetown, MA 02657
Phone: (508) 487-3622
Fax: (508) 487-4495
http://www.provincetown.com/coastalstudies/index.html
ccswhale@wn.net

Provincetown is one of the nicest little towns in the country populated by some of the nicest people you'd care to meet. It becomes a big tourist and vacation hotspot in the summer, when the population more than triples, and housing becomes scarce. There are dozens of historic structures scattered around Cape Cod, which is crawling with sandy dunes and shell-strewn beaches. The leeward side of the peninsula not only provides shelter for small boats and swimmers, accumulates stones and pebbles of a thousand different hues. The town has an historic feel, dozens of touristy shops and restaurants, and an old and strong connection to the marinelife which literally surround it.

Overview

The Center for Coastal Studies is one of the primary lines of defense for the marine ecosystems of Cape Cod and Massachusetts Bay, and for the animals which inhabit them. For over 20 years the Center has been collecting and analyzing information about local water quality, toxic compounds, habitat conditions, and how these factors affect the marinelife which uses them. Center scientists work with government researchers and prestigious private institutions like Woods Hole Oceanographic Institution. They map out the movements, sizes, and dynamics of many marine mammal populations, including minke whales, fin whales, humpback whales, and the most endangered of the large whales, the North Atlantic Right Whale.

Conservation efforts by the Center succeed in finding new uses for discarded drift nets, erecting natural history displays along Cape Cod's shorelines, and reducing the number of entangled harbor porpoises in gill nets by convincing local fishermen to use acoustical devices pingers on their gillnets. The Center exposes hundreds of local residents and thousands of annual visitors to a wealth of natural history and conservation facts and issues through whalewatching, Elderhostel, school programs, and its Summer

Institute, which offers courses for college and graduate credit.

The center is the only group with federal authorization to disentangle whales which become trapped in fishing nets. The Center's Rapid Response Rescue Program, developed in conjunction with the Coast Guard, is the first of its kind. The Center's rescue team is also responsible for rehabilitation and release efforts during incidents of mass pilot whale strandings along Cape Cod.

(Photo provided by The Center For Coastal Studies156)

Visiting the Center for Coastal Studies gave me the distinct impression that someone named Mabel was going to materialize at any moment with a platter of biscuits.

Research Projects

There are several cetacean-related research efforts underway at the Center, each managed by a primary researcher. A behavioral and habitat study under the direction of Dr. Charles Mayo has been investigating the population and habitat use of North Atlantic right whales for some time. Most field research for this study is done between the months of January and May, making the competition from college students less intense. Dr. Moira Brown is conducting population biology studies and a comprehensive genetic analysis of right whales, working in both U.S. and Canadian waters. In the summer months another Center staff researcher, David Mattila, heads out to investigate the populations of humpback whales in the Gulf of Maine and surrounding waters. Yet another researcher, Margaret Murphy, has taken on the task of cataloguing minke whales in the North Atlantic, a task made difficult by the whales'

(Photo provided by The Center For Coastal Studies<None>)

CCS research isn't limited to cetacean work. Center scientists study debris, pollution, and many other areas of marine conservation.

(Photo provided by The Center For Coastal Studies<None>)

Here a staff member videotapes a scientist taking notes on a photographer who is documenting a researcher conducting a study on whales.

small size, reclusive tendencies, and the lack of any prior photographic catalogue.

Involvement Opportunities

While the Center for Coastal Studies isn't the sort of place where one hopes to get a permanent job (there just aren't very many available) it is one of the best places an aspiring cetacean scientist can go for experience. The Center makes use of both volunteers and interns for a wide variety of tasks, all of which provide exposure to some of the country's most experienced marine mammal researchers.

Volunteers and interns can usually get involved in any of the Center's programs. Many projects are seasonal, depending on the habits of the animals being studied, so a little advance planning is in order. (Six months to a year would be appropriate.) Participants can expect to have their hands full recording data, identifying, and cataloguing photographs, and cleaning equipment. Volunteers and interns are often brought along aboard research vessels to assist with sightings.

Volunteers assist in virtually every aspect of the Center's operations, including clerical work, data entry, whale observations, and photo ID work. Individuals wishing to assist in whale research must be able to commit to at least two months of service, so the Center has adequate time to make use of its training program.

The Center's Summer Institute offers intensive courses in marine and coastal science, combining classroom, field, and laboratory studies. Participants can earn undergraduate, graduate, or professional development credits for courses.

The Center helps provide low-cost housing for interns through cooperative agreements with the local arts center and a nearby youth hostel. All things considered, it's definitely one of the best deals anywhere. The duration of an internship at the Center is flexible, but usually corresponds to the student's academic semester. Interns take on a particular project to complete, usually something related to one of the Center's ongoing research programs. Applications for summer internships should be submitted no later than February 1st.

Although the Center does use volunteer assistance with stranding response, interested parties should contact the International Wildlife Coalition (entry on page 235) which handles volunteer screening and training for the Cape Cod Stranding Network.

Other Programs

Members receive an annual subscription to the Center's newsletter, Coastwatch, which does an excellent job of updating members on current Center projects and issues. Members receive discounts on whale watch trips and other events, free admission to lectures and field walks, and access to the Center's library of marine-related books and materials. Memberships start at $20 for individuals.

Center For Whale Research

P.O. Box 1577
Friday Harbor, WA 98250
Phone: (360) 378-5835
Fax: (360) 378-5954
Kennith C. Balcomb III - Executive Director

Overview & Research Projects

The Center for Whale Research is based in Friday Harbor, Washington. It is a nonprofit, all-volunteer group whose primary goal and function is to conduct long-term photo identification studies on orcas in the Pacific Northwest. In 1973 a Canadian researcher figured out that all Orcas have unique and distinctive markings and dorsal fins, and that by photographing the whales when they surface he could keep track of who was who without using invasive methods. The Researcher in question was Mike Brigg, who in some circles is called the "Father of killer whale research." Although he passed away in 1995, his name lives on: orca #26 in "J" pod was named "Mike" in his honor.

The Center's research is critical in providing the scientific community with the data required to conduct effective studies on wild cetaceans. It's also a heck of a lot of work, and these guys certainly have my respect. Imagine having to sit on a street corner for years until you can tell who each passing pedestrian is by freckles on their faces. That's how accurate the Center's data is on the dorsal markings of the more than three hundred Orcas living off the Pacific Northwest coast.

In addition to mapping out the Orcas in the northern Pacific, the Center is also running similar photo identification studies on Pacific humpback whales, as well as a general survey of marine mammals in the Bahama islands, focusing on bottlenose dolphins, pilot whales, and beaked whales.

Involvement Opportunities

If you'd like to become involved in these studies you should contact Earthwatch (see separate entry) as all Center for Whale Research volunteers are handled through their office as expedition members. Returning volunteers may become volunteer staff, but have to go through Earthwatch initially. A typical ten day session costs about $1800 including room and board.

Other Programs

Memberships are a bargain at $20/year, and include a subscription to the biquarterly report Orca Log, and periodic newsletters. Orca Log provides current information about cetaceans, research projects, and events in the marine mammal scientific community.

The Center for Whale Research does not condone "show business captivity" of cetaceans, and is active with other organizations in the scientific evaluation of the possibility of returning selected animals to the wild. The Center offers two publications on the issue, Cetacean Releases and An Annotated Bibliography on Cetacean Releases, both for $10 each, including shipping costs.

Cetacean Behavior Laboratory

Psychology Department
San Diego State University
San Diego, CA 92182
Phone: (619) 594-5649
Fax: (619) 594-1332
http:www.sci.sdsu.edu/cbl/cblhome.html
Dr. R.H. Defran - Director
rdefran@sunstroke.sdsu.edu

Overview

The CBL's Director, R.H. Defran, has been a professor in the Psychology Department at San Diego State University for almost thirty years. He has taught modern training techniques to both Sea World and Naval marine mammal trainers, and compiled a volume on the social ecology of eleven cetacean species. Defran was approached by the National Marine Fisheries Service in 1983 with an offer to assume control of a photo identification project along the Southern California coastline. Defran and the CBL have continued the study ever since, expanding its range, collaborating with other researchers, and conducting behavioral research as well.

Research Projects

The CBL's efforts mapped out a large portion of the resident bottlenose dolphin populations along the Southern California coastline, as well as along the Mexican Baja peninsula. CBL researches the dolphins' behavior and social structure, primarily through land-based observations. The Lab has tried to discover the relationship of such factors as time of day, tidal state, ocean and kelp condition, wave height, water depth, and season to the dolphins' behavior.

Involvement Opportunities

Opportunities for graduate study at CBL are open to students who are committed to pursuing a Ph.D. in cetacean science. While CBL itself can only offer a Master's program, it can provide a student with the credentials needed for admission to a Ph.D. program in marine mammal science at another facility. CBL gives preference to students with a strong academic record, field research experience with photography, small boat handling, and experience with Macintosh and PC computers. Students with little field experience must complete six months of training before officially joining the program.

Interns at CBL spend a three month summer or semester working and being trained in cetacean field work and data analysis, and are required to review a fair amount of cetacean research literature. Participants work about twenty hours per week, which is spent both in the laboratory and in the field. Field work consists mostly of land-based observations made from oceanside clifftops, though interns occasionally get to assist with boat-based research as well. In the lab, the interns learn to use spreadsheets and other applications in data analysis, focusing on the home range, travel, feeding, socializing, playing and resting behavior of the dolphins.

Summer internships run from June through August, fall internships from September through December, and Spring internships from February through May. The program is open to advanced undergraduate or recent graduate students in the behavioral or biological sciences. Participants must pay a fee of $750 which does not cover room, board, or transportation. Applicants should send a resume and cover letter, preferably by Email, which includes a brief biographical sketch, a detailed academic history, and a listing of any relevant skills or interests. The letter should indicate a desired semester for the internship, as well as any existing time constraints. Letters should be 2 to 3 pages, single spaced and sent directly to Dr. Defran, along with academic transcripts and two letters of recommendation, written by faculty members familiar with the applicant's work.

Cetacean Research Unit

P.O. Box 159
Gloucester, MA 09130
Phone: (978) 281-6351
Fax: (978) 281-5666
http://www.friend.ly.net/user-homepages/b/birdman/index.htm
info@cetacean.org
Dave Morin - Intern Coordinator

Overview

The Cetacean Research Unit was formed in 1979 as a vehicle for whale research and conservation in New England. Working in tandem with a host of scientists from colleges and universities throughout the east coast, the CRU has tackled dozens of individual research projects ranging from population studies to toxins and bioacoustics. The CRU provides the results of its research to the scientific community through published reports, and also seeks to educate the public through its membership journal and educational outreach programs. The CRU is very active in marine conservation efforts, and was instrumental in the formation of the Stellwagen Bank National Marine Sanctuary in 1993. CRU is a co-leader of the Coastal Advocacy Network, a coalition of over 150,000 individuals and groups working to protect the New England Coastline.

Research Projects

The CRU identifies and catalogs humpback whales in the New England area and has identified over one thousand individuals. The Unit initiated studies of right, blue, fin, minke, and sei whales, as well as Atlantic white-sided dolphins. The CRU uses three methods of data collection for these studies. Whale-watching trips provide them with broad exposure to the animals all along the coastline. The CRU owns a small research vessel of its own, which it uses for more comprehensive studies. It maintains a network of whale-watch naturalist observers throughout New England, who record and report observations to CRU scientists.

Involvement Opportunities

CRU uses no volunteers, but does offer internships to ten upperclass college students and recent graduates each year.

Participants pay about $1,000 to cover housing and food. The internship lasts for an entire semester, during which the participant becomes familiar with the Unit's research programs and techniques, and participates in conservation efforts. Certain applicants may be awarded grant money to defray the expense, when available. Applicants should have a background in biology, some knowledge of research methods, good teamwork skills, and the dedication to work long days in the field or laboratory.

Summer is the most active season for CRU research, making the four summer internships the most coveted. The remaining six internships are divided evenly between the fall and spring seasons. Exceptional interns have been known to return to CRU to join the staff as naturalists and researchers. Applications are taken between January 1st and March 1st for the following year, and should include a resume, cover letter, transcript, and one letter of recommendation.

(Photo provided by CERF)

Coastal Ecosystems Research Foundation (description to the right) carries out population studies on gray whales.

Coastal Ecosystems Research Foundation

c/o Adventure Spirit Travel Company
1843 W. 12th
Vancouver, BC, Canada
V6J 2E7
Phone: (604) 736-5188
Fax: (604) 732-0476
megill@zoology.ubc.ca
http://www.zoology.ubc.ca/~megill/cerf

Overview

The Coastal Ecosystems Research Foundation is a nonprofit organization which offers research expeditions to the paying public. The CERF's research efforts focus on the biology, ecology, and conservation of dolphins and whales off the western coast of Canada. CERF is a relative newcomer to the public research expedition business, and has shown promising signs of growth over the past few years.

Research Projects

While the schedule and research focus for CERF's 1998 schedule was not available at press time, the seven-day research expeditions are likely to be run from late June through early September. CERF spent the 1997 season studying the habitat use, interaction, and feeding habits of the local resident grey whales. Each excursion took place aboard the organization's 40′ sail craft, amid the spectacular scenery of British Columbia's Central Coast.

The 1998 CERF research season will run from late June through early September, and will focus on gray whale, humpback whale, and Pacific white-sided dolphin distribution, abundance, and feeding ecology in Queen Charlotte sound and neighboring waters.

Involvement Opportunities

Expeditions are open to anyone who is interested. Participants get involved in data collection and analysis, as well as photo identification efforts. Costs for the week-long trip generally run at just over $1000 U.S., with a slight discount for students. Food and transportation to and from Port Hardy, Canada is provided, and accommodations are provided in the form of camping tents along the shore. Evening lectures on various cetacean research topics are offered four nights per week, as well as short discussions of

The Dolphin Project

P.O. Box 10323
Savannah, GA 31412
Phone: (770) 426-4902
Phone: (912) 925-7420
http://www.lads.com/TheDolphinProject

Overview

Following incidents of massive bottlenose dolphin dieoffs in 1987 and 1988, a consensus was easily formed amongst Eastern Seaboard researchers that a great need for basic information about these animals still existed. The Dolphin Project was formed to help accomplish that task, and has been working ever since to map out the populations and behavior of bottlenose dolphins within the barrier islands region of Georgia and southern South Carolina. The Project is an all-volunteer outfit which conducts quarterly surveys aboard small vessels to record the animals' abundance within a 100 mile stretch of coastline. The program collaborates with some of the most respected cetacean scientists in the country, and places an emphasis on effective training, and the acquisition of scientifically valid and usable data.

Research Projects

The Project conducts two types of surveys which take place in 56 designated zones. Abundance Surveys are carried out four times per year, and take place over a weekend. Two surveys are done on Saturday, and a third on Sunday morning. Each zone's surveys are carried out by a four-person crew, consisting of a team leader, skipper, data recorder, and a photographer. During the Abundance Surveys, particular attention is given to recording data about the number and distribution of animals encountered, and photographs are taken on an opportunistic basis.

Occurring seven times each year, special Photographic Surveys allow TDP to keep track of individual dolphins. Also consisting of a four-person crew, the data recorder is replaced with a videographer during a Photo Survey, the main purpose of which is to record usable dorsal fin images and behavioral video footage.

Involvement Opportunities

TDP uses almost 300 volunteers for its surveys, nearly all of whom are Georgia and South Carolina residents with little or no prior scientific experience. The Project has a small but ever-increasing number of participants from Florida as well. The Project has plans to open a Jacksonville, FL chapter in 1998, and may begin to survey Florida waters if membership continues to grow in that region, and if collaborative arrangements can be made with local government and academic groups.

Volunteers are required to undergo training before participating in an Abundance Survey as a photographer or data recorder. Training workshops are open to members of the project, and take place for a half-day on each Saturday morning prior to an abundance survey. Reservations are not required, though you must contact TDP for locations and times.

Volunteers may also take an additional training course to become eligible to participate in the Photo Surveys, which are allowed by special permission from the government to approach the animals rather closely. Volunteer assistance is also needed to help catalogue photos and sort data.The additional training for the photographic surveys takes a mere hour's time, and takes place at a college in Savannah the night prior to each trip.

Memberships are available for as little as $25, or $19 for students. Once trained, members may sign up to participate in an Abundance Survey, and are required to pay a small fee to offset operational costs. Here's the 1988 schedule for TDP's training and surveys:

Training: (Savannah and Atlanta) December 6th, (1997), March 7th, June 6th, September 12th, and December 5th.

Abundance Surveys: January 24th & 25th, April 18th & 19th, July 11th & 12th, October 17th & 18th, and January 23rd & 24th (1999).

Earthwatch Institute

680 Mt. Aubern St.
P.O. Box 9104
Watertown, MA 02272
Phone: (800) 776-0188
Fax: (617) 926-8532
http://www.earthwatch.org
info@earthwatch.org

Okay, so you want to play Indiana Jones, but only for a week, and no hostile natives or monkey brain stew? Earthwatch may be just what you're looking for. For more than twenty five years Earthwatch Institute has been supplying field researchers throughout the globe with eager volunteers, and has given those volunteers invaluable experience with field studies and exposure to some of the world's leading research scientists.

Overview & Involvement Opportunities

Earthwatch expeditions cover ornithology, spelunking, Russian folklore, reef surveys, paleontology, and dozens of other areas of biological, anthropological, and geological research. Earthwatch even offers opportunities for cetacean enthusiasts. Upcoming trips will focus on sperm and humpback whales, bottlenose and dusky dolphins, and orcas. Earthwatch volunteers are given a unique opportunity to jump right into the heart of some of the world's premiere cetacean research programs with little or no prior training or experience.

The catch is you have to pay for it.

Earthwatch expeditions are very pricey to run, and each participant carries their fair share of the burden by making a tax-deductible contribution toward project costs. A ten day stint will probably cost over fifteen hundred dollars, and some of the two-week deals can set you back a full two bills. An Earthwatch trip really is a chance to join a working field research team, so accommodations can be rather rural and Spartan, depending on the nature of the project. Earthwatch has a long history of satisfied volunteers and responsible work, and many participants finish their experience even more satisfied with the work they accomplish than with the amazing sights they behold. (And there's a certain high to watching a NOVA special on wild dolphins and knowing you helped make that knowledge possible.)

An advance contribution (a $250 deposit) is required to reserve one of the often limited spaces on Earthwatch teams, which is fully refundable until 90 days prior to departure, and partially refundable after that point. The cost of the project doesn't cover airfare to the rendezvous point, but it does cover pretty much everything else, and Earthwatch will be more than happy to point you in the right direction for reasonable airfare rates. Both students and teachers may qualify for a limited number of grants offered through Earthwatch for their expeditions (over 300 participants were funded last year by corporations and foundations). Contact the main office for more details.

Earthwatch occasionally has openings for paid staff, though these positions are usually clerical or administrative in nature. Job listings are maintained on their web site at

http://www.earthwatch.org/t/Temployment.html.

Below are Earthwatch's descriptions of their cetacean-related expeditions for 1998, though I would strongly advise anyone considering them to grab a copy of their catalog or sign on to their web site and check out the other expeditions as well.

Research Projects

Australian Humpbacks
Project leaders:
Curt Jenner, Micheline-Nicole Jenner and Kenneth Balcomb
Center for Whale Research

Buccaneer Archipelago, Western Australia: Although humpbacks off Australia have increased in numbers over the past few years, centuries of whaling have cut the world humpback population to a mere fraction of its former size. Major migration routes stand as the some of the few places where we can take a rare glimpse into the mysterious life of humpbacks and better understand how these enormous and acrobatic marine mammals behave and what they need to survive as a species. Since 1990, the Jenners have been observing whales and whale behavior on one major migration route off the northwest coast of Australia. For the seventh season out of an eleven-season study, the first long-term

EARTHWATCH EXPEDITIONS, CONT.

humpback study in the area, EarthCorps teams will help the Jenners and Ken Balcomb, Principal Investigator of Orca, observe, identify, and census migrating humpbacks. The primary objective of this decade-long study is to determine the location of the humpback's calving grounds. Do all the humpbacks following this route use one discreet area to bear their young? The second objective is to refine the world humpback population estimate. How much have numbers recovered since the 1960s? The third goal is to determine the relationship between the northwest Australian whales and other humpback populations in the Southern Hemisphere. Do they interbreed or even intermingle? From aboard a 12-meter (40-foot) catamaran called the RV WhaleSong, teams will address these issues and also try to determine where these migrants are heading.

Bahamas Marine Mammal Survey
Ken Balcomb, III & Diane Claridge
Center for Whale Research

Abaco Cay, Bahamas: Although we are not sure exactly how large marine mammals use the Caribbean, local Bahamians know that the warm, clear waters are full of a variety of cetaceans. Spotted dolphins, bottlenose dolphins, beaked whales, pilot whales, sperm whales, and humpback whales are still spotted between the tiny islands despite decades of commercial hunting. Volunteers are needed to inventory and examine this sea-mammal-Mecca so that we can better understand our marine counterparts. Although whales, dolphins, and seals have been known to exist in Bahamian waters for some time, their abundance, seasonality, and distribution are still largely unknown. Ken Balcomb and Diane Claridge, with help from a fourth season of EarthCorps volunteers, will conduct an inventory of marine mammals and their habitats in the Bahamas. By conducting visual and acoustic surveys, teams will locate areas frequented by marine mammals. The primary objective of the project is to further establish migratory routes, so that scientists can conduct long-term investigations. The goals of the longer investigations are to examine the nature of these Caribbean whale and dolphin populations. Are they residents? If they're just passing through where do they come from and how long do they stay? Are they related to populations in other regions

EARTHWATCH EXPEDITIONS, CONT.

of the Atlantic? Only through evaluating the size and dynamics of cetacean populations can we begin to establish effective conservation measures.

Exploring Dolphin Intelligence
Drs. Louis Herman & Adam Pack
Kewalo Basin Marine Mammal Laboratory, University of Hawaii

Honolulu, Hawaii: At first glance, a dolphin may seem to possess at least a glimmer of intelligence in its dark eyes. After a dolphin splashes an unwitting spectator, you may swear it is laughing. But a dolphin can also remember, communicate, and to varying levels, understand. For centuries, humans have acknowledged that dolphins are indeed intelligent. For a 15th year, Earthwatch volunteers will work to discover just how intelligent they are. Staged at the University of Hawaii's Kewalo Basin Marine Mammal Laboratory, this extremely popular project is a long-term investigation of the cognitive, intellectual and communicative capabilities of bottlenose dolphins. Large, outdoor, seawater tanks are home to four bottlenose dolphins that are engaged in a comprehensive education program. Each dolphin builds up a knowledge base of problem-solving strategies and concepts that ultimately allows the researcher to ask increasingly complex questions. Ongoing studies include investigations of memory, sensory processes, and capabilities for understanding artificial languages. Phoenix and Akeakamai are two 18-year-old females that have received extensive training in accepting instructions and responding to questions expressed with artificial gestural and acoustic languages. Hiapo and Elele, a 9-year-old male and female, respectively, have been the subjects of numerous conceptual studies that have tested their ability to integrate visual and echolocation information. Typically, each dolphin participates in several studies throughout the day or week, and different dolphins are often involved in different studies. This strategy encourages increased research productivity and helps to maintain the interest and motivation of each dolphin. This year, EarthCorps volunteers will continue to address dolphin intelligence by trying to answer a series of questions. How does a dolphin recognize and remember? Do dolphins depend more on vision or sound? Do dolphins understand analogies? To what extent do dolphins "speak"? What role does mimicry play in dol-

EARTHWATCH EXPEDITIONS, CONT.

phin communication? How do dolphins mentally represent their world? To what extent do dolphins understand language?

Humpbacks off Hawaii
Dr. Louis Herman and Dr. Adam Pack
Kewalo Basin Marine Mammal Laboratory, University of Hawaii at Manoa

Maui, Hawaii: As humpback whales travel their seasonal migrations, whale watchers and scientists are provided excellent opportunities to observe these giants in action. Regularly swimming from their northern summer feeding grounds to southern breeding grounds, humpbacks move slowly and jump almost fully out of the water, displaying to onlookers the full magnificence of their often 15-meter (50-foot) bodies. Unfortunately, it is this same predictability and playfulness that has made humpbacks easy targets for mass slaughter. Because the Hawaiian Islands are the major reproductive grounds for North Pacific humpbacks, the area is ideal for better understanding this dramatically depleted species. The Kewalo Basin Marine Mammal Laboratory has been conducting a long-term study of humpback whales in the North Pacific since 1975. For the past 20 years, they have observed humpback migration, reproductive histories, and communication and have compiled an enormous volume of tail-fluke photographs for population surveys. In addition to assisting in this long-term photo-identification program, 1996 teams will concentrate on studying humpback behavior. From the shores and waters off Maui, teams will complement water-surface humpback observations with below-the-surface data. How do the whales use their songs? How do the whales interact during breeding season? Through better understanding the needs of these whales, scientists at Kewalo are developing sound management plans. Research Area

Orca
Kenneth Balcomb III & Dr. Astrid Maria van Ginneken
Center for Whale Research

San Juan Island, Washington: In 1862, when D.F. Eschricht reported finding the remains of 14 seals and 13 porpoises in the stomach of one orca floating off the coast of Jutland in the North

EARTHWATCH EXPEDITIONS, CONT.

Sea, he bolstered public opinion that orcas are killers and mindless, shark-like, eating machines. Over time, this image changed, and captive orcas became popular exhibits at oceanariums. Ironically, the more scientists began to realize the intelligence and individual character of these marine mammals, the more orcas became victims to a lifetime of marine-park captivity. The captures ceased in 1976, and though remaining populations were small, we have been able to learn a considerable amount about orca ecology from the small Pacific Northwest population. Some orca communities thrive on salmon that occupy the area while others feed exclusively on seals. This diversity of behavior seems to indicate that orcas live in complex societies and learn through the habits of their individual groups. From historic whaling and stranding data, we have known for a few decades that, in some parts of the world, these animals grow to be about 7 to 10 meters (23 to 33 feet) in length, are sexually mature at around 5 meters (17 feet), and that males can weigh up to 39,600 kilograms (9 tons). Despite our accumulated data, we cannot be sure whether the Puget Sound orca population or its ecosystem is viable in the face of increasing human activities that pollute, damage, and consume the resources; salmon are one of the several marine species which are in danger in the Pacific Northwest as a result of over-fishing and human river development. Their passing to extinction does not bode well for the predators, including humans. Besides accumulating general orca knowledge, our immediate goals are to generate publicity that can spur the necessary support for reducing these damaging activities.

Whales of South Africa
Dr. Peter Best, Mammal Research Institute
South Africa Museum

Walker Bay, St. Sebastian Bay, South Africa: At the turn of the century, right whales were deemed the "right" whale to hunt by whalers because of their low speeds, buoyancy after death, and valuable oil and baleen. After 7,000 whales were taken from the east coast of Africa between 1910 and 1915 and over 4,000 whales were taken from Madagascar waters between 1937 and 1939, whalers began to acknowledge that catch rates were sharply declining. The threat of extinction eventually prompted worldwide protection of the humpback in 1963, and since 1979, the Indian

EARTHWATCH EXPEDITIONS, CONT.

Ocean north of 55ûS has been declared a whale sanctuary by the International Whaling Commission. Because of the rare opportunity to witness a possible flourishing of whale populations, the Mammal Research Institute (MRI) began shore-based surveys at Cape Vidal, Natal, in 1988. Starting in 1991, with the help of Earthwatch, the MRI has carried out whale-spotting cruises in southern Mozambique and Madagascar. According to Dr. Peter Best, the area's right whale population appears to be increasing at a rate of 7 percent a year. For four field seasons, the objectives of this first-ever systematic whale survey in the area have been to assess the distribution and size of the humpback whale population visiting southern Madagascar and Mozambique in winter. Ultimately, teams will work to determine the current dynamics of the humpback populations in this Indian Ocean sanctuary. What factors contribute to the increasing numbers of whales? Through determining evidence of breeding such as whale songs, interacting groups, and cow-calf pairs and population changes, crews will analyze recruitment rates and the mechanisms of reproduction and growth.

Wild Dolphin Societies
Dr. Randall Wells, Chicago Zoological Society
Mote Marine Lab.

Sarasota Bay, Florida: For thousands of years, humans have been fascinated by dolphins. Even before the first oceanariums began displaying captured dolphins in the beginning of this century, people were aware of the almost human-like intelligence exhibited in the graceful sea mammals. Today, we are still learning about the incredible range of cognition, acoustical communication, and complex behavior enabled by their large brains. Beyond the obvious trainability of dolphins, they demonstrate creative understanding through identifying themselves with distinctive whistles. Although an individual's whistle generally resembles that of its mother's, each dolphin's trademark is as easily identifiable as a advertising jingle. As human activities increasingly destroy dolphin habitats, we have an urgent need to better understand our marine counterparts. The coastal nature of bottlenose dolphins makes them particularly susceptible to these impacts. In recent years dolphin populations have been impacted by fisheries, oil spills, vessel traffic,

EARTHWATCH EXPEDITIONS, CONT.

and exposure to high concentrations of toxic pollution. Scientists need more accurate and detailed information to combat these threats. Because biological data from free-ranging dolphin populations is obviously difficult to collect, Randy Wells has spent the past 25 years, 13 of which with EarthCorps volunteers, capturing, testing, marking, and releasing dolphins. Though teams will not capture dolphins in the upcoming field season, markings made in past years will help teams identify individuals. Volunteers will continue the long-term study by examining and recording the behavior, interaction, population biology, and health of Sarasota Bay's resident community of 100-odd bottlenose dolphins. Teams will focus specifically on seasonal habitat and activity changes.

New Zealand Dolphins
Suzanne Yin & Dr. Bernd Wursig
Texas A&M University

Kaikoura, New Zealand: On any given morning, up to 300 dusky dolphins might be spinning and somersaulting near the coast of New Zealand's Kaikoura Peninsula. Because dolphins are by nature friendly, not to mention extremely acrobatic and entertaining, they have become the area's major tourist attraction. Volunteers will observe human-dolphin interaction in New Zealand to determine how tourism affects dolphin communities. In the dark of night, dusky dolphins generally feed in deep waters further than three kilometers (two miles) from shore. When the sun comes up, the marine mammals head for the warm, shark-free coastal flats. During these daytime play periods, dolphins rest, casually feed, and seem to teach their young habits for survival. Tourist companies in Kaikoura take advantage of these opportunities to send boatloads of people to watch and swim with the dolphins. Although the tour operators are generally responsible and the dolphins do not appear outwardly disturbed, we are not sure how this interaction affects the schools. Without disturbing the dolphins, Earthwatch volunteers will observe and document interaction between humans and dusky dolphins in the near-shore waters of South Island. Do human activities alter the dolphins' behavior and movement patterns? Do the dolphins respond differently to different human activities? How can we help direct and manage tourist activities? We hope that this study will help inform dolphin-

EARTHWATCH EXPEDITIONS, CONT.

tourist operations worldwide.

Australian Dolphins
Dr. Peter J. Corkeron
James Cook University

North Stradbroke Island and Moreton Bay, Eastern Australia: Bottlenose dolphins are among the most studied of the cetaceans and occur along the entire Australia coastline. Between 1984 and 1987, over 330 individual bottlenose dolphins were identified in Moreton Bay from photographs of natural marks. In addition, 321 individual dolphins were identified in the inshore oceanic waters around Point Lookout, North Stradbroke Island. The overall objective of this project is to assess the level of anthropogenic impact on bottlenose dolphins, Tursiops truncatus, in Moreton Bay by quantifying the ranging patterns and behavioral time budgets of individual dolphins in the Bay and an adjoining area, Point Lookout. Specific objectives include: attaching radio transmitters and MST-VHFs to bottlenose dolphins in the Bay and in the waters off Point Lookout; determine the ranging and diving patterns, and behavioral time budgets of bottlenose dolphins in the waters off Point Lookout by triangulation of radio signals received from shore stations; determine the ranging and diving patterns and behavioral time budgets of dolphins in Moreton Bay by a) surveys to locate radio-tagged animals and b) focal follows of tagged animals.

Grey Whales Off Baja
Francisco Ollervides & Dr. William Evans,
Texas A&M University

Pacific Baja, Mexico: One of the best places you can watch grey whales and their calves up close is the Boca de Soledad ("Mouth of Solitude") lagoon on the Pacific coast of Baja California. In 1994, more than 10,000 tourists viewed the whales here aboard locally operated tour boats. Although Bahía Magdelena is not the biggest grey-whale breeding and calving area on Baja, the whales are more concentrated in this shallow, narrow lagoon than elsewhere. So cow whales with calves may feel more vulnerable to gawking tourists and to noisy, exhaust-spewing, whale-watching boats. And, unlike the two larger grey-whale breeding hotspots on Baja, Bahía Magdelena is not legally protected; hence its booming whale-

EARTHWATCH EXPEDITIONS, CONT.

watching industry is unregulated. Although the thriving eastern Pacific stock of grey whales was removed from the U.S. Endangered Species list in 1994, the impoverished Bahía communities who depend on the whales don't want to kill the goose that laid the golden egg, particularly since Bahía Magdelena is a cradle for the next generation of grey whales. How can the local fishing villages ensure that their livelihood does not threaten the whales' welfare? Working with Francisco Ollervides, a masters candidate at Texas A&M University, you'll help assess the impact of boat traffic on the whales and contribute data directly to setting up guidelines for Bahía Magdelena's whale-watching industry - in the best interests of both the boat operators and the whales. The focus of the research is monitoring sounds (from boats, whales, ambient noise, and any other sources) and correlating the whales' behavior (especially diving and breathing rates and movements) with the sounds and boat positions. That requires a lot of people on watch at the same time, all linked with two-way radios.

Baja Sperm Whales
Nathalie Jaquet, Ph.D.
Otago University, New Zealand

Gulf of California, Baja California Sur: The calf shadows his mother, who's blowing seven times a minute. Sending spray eight meters into the air, she's preparing to dive a kilometer or more for squid in ocean trenches. Squid breath lingers in the air. You maneuver the boat into position; a teammate shoots length-measurement photos with a 300-mm lens. Then the female dives, her finless back slowly rolling forward and fluked tail arcing gracefully. The crew snaps identification shots of the flukes' unique skin patterns. When the boat arrives at the smooth whale slick, marking where she dove, your team mates scoop up squid beaks she left behind. The beaks are --Dosidicus gigas --the target of the vibrant and unregulated local squid industry.

That industry is part of why you're here with Pavia and Dr. Nathalie Jaquet, a Swiss marine ecologist who pursued her doctoral research on Pacific sperm whales under renowned cetacean specialist Hal Whitehead at Dalhousie University. Although sperm whales (with an estimated world-wide population of just under two million) are the most abundant of the seven great whale

EARTHWATCH EXPEDITIONS, CONT.

species, their survival is by no means assured. For one, their reproductive rate is appallingly slow, in part because the selective slaughter of males (for their larger amounts of valuable oil) upset the social and reproductive balance. Moreover, scientists are concerned that commercial squid fishermen may be depleting the sperm whales' main food supply. (A mature male may eat a ton of squid per day.) Today, these deep-diving creatures of the open ocean, the largest toothed cetaceans, remain the most enigmatic--and most challenging to study--of all the great whales. "Above all other hunted whales, his is an unwritten life," Herman Melville wrote of the sperm whale. To write that life, Jaquet has enlisted your help to survey, identify, and track the Gulf of California sperm whales, sample water quality, and test if patches of squid determine sperm whale distribution. You're in good hands. Beyond her oceanographic experience, Jaquet, like Pavia, also has a skipper's license and extensive sailing experience.

Hubbs-Sea World Research Institute

2595 Ingraham St.
San Diego, CA 92109
Phone: (619) 226-3870
Fax: (619) 226-3944

Overview

Hubbs-Sea World Institute is a nonprofit marine research institute supported by, you guessed it, Sea World. The Institute sits adjacent to Sea World of California in San Diego, overlooking cozy Perez Cove, which in turn is nestled in the sparkling waters of Mission Bay. The sky is always blue and nobody ever talks about the weather because it's always gorgeous. (I'm writing this a mere two feet away from the throes of a Chicago snowstorm which is currently raging on the other side of my single-paned window. Grr.) All this is just east of the hip-happenin' Mission Beach district, where everyone plays volleyball and looks like they're about to audition for a Charles Atlas ad. The land under the institute is owned by Sea World, who charges them exactly one dollar a year for rent, which is about the best deal you'll find in San Diego.

Research Projects

The Institute conducts a number of research programs aimed at advancing scientific understanding of aquaculture, bioacoustics, physiology, ecology, and marine conservation. Results from the Institute's studies are used to provide information needed for environmental management decisions and conservation programs. The Institute's Bioacoustics laboratory has been involved with recent world-wide efforts to determine how various man-made noises affect marine mammals, including certain portions of the ATOC study (see separate entry). Research biologists and associates at the Institute have conducted a number of studies involving cetaceans at Sea World, often with the assistance of students from local universities like the University of San Diego and San Diego State University. The Institute has also been involved with developing satellite tracking packages for use with wild cetaceans.

There is a smaller counterpart facility in Florida (working in conjunction with the Sea World facility in Orlando) which has studied local dolphin feeding habits, gathered life history data on the relatively unknown pygmy sperm whales which often strand on

Florida's east coast, and which maintains a database on the dolphins and whales which strand within the Southeastern Marine Mammal Stranding Network. Send inquiries about the Institute's Florida activities to Dr. Dan Odell at Sea World of Florida (see address on page 127.)

Involvement Opportunities

Hubbs-Sea World is a small institution, and opportunities to get involved in their work are very rare, though they do exist. The easiest way to do it would probably be to take a class at one of the schools mentioned above which is being taught by a Hubbs-Sea World researcher, and shmooze your way in from there. Ask the department chair in the appropriate school (or whoever takes calls for them) if they are aware of such a class. (If you see one being taught by Dr. Ann Bowles, snag a seat.) Most students who have gotten involved in the Institute's research have been graduate students, but a very few undergrads have snuck in too.

Intersea Foundation

P.O. Box 1106
Carmel Valley, CA 93924
Phone: (408) 659-5807
Fax: (408) 659-5821
http://www.intersea-fdn.com
whale@intersea-fdn.com

Overview

The Intersea Foundation offers the public opportunities to participate in research expeditions, primarily in Southeast Alaska. These trips cater to aspiring scientists looking for exposure to field research techniques, as well as to vacationers hoping for a more relaxed look at wildlife. It was founded in 1976 by Cynthia D'Vincent, a well known marine biologist and photographer, and the scientist responsible for the discovery of cooperative feeding among North Pacific humpback whales. The Foundation conducts marine research programs in Alaska, California, Hawaii, Mexico, the Marquesas Islands, and the Bering and Beaufort Seas.

Research Projects

The Foundation's research efforts focus on humpback whales, including photo identification, acoustic analysis, and behavioral studies. Expeditions are offered in conjunction with California State University, Monterey Bay. The trips are usually just over a week long, and take place in the protected waters of Alaska's Inside Passage, where orcas, minke whales, Dall's and harbor porpoises, and various pinnipeds can be found, in addition to the humpbacks.

Involvement Opportunities

Primary involvement in the Foundation's efforts is through the public research expeditions. Final schedules and costs for 1998 had not been set by press time, but trips usually run between late June and early September, and cost around $2,500 per person. Contact the organization directly for a brochure, or check their website for updates.

Other Programs

The Foundation periodically offers lectures in California, and occasionally in other states as well. Again, contact the Foundation for details.

Marine Mammal Research Program

Behavioral Ecology Division
Texas A&M University
4700 Avenue U, Bldg 303
Galveston, TX 77551
Phone: (409) 740-4718
Fax: (409) 740-4717
http://www.tamug.tamu.edu/mmrp
Dr. Bernd Wursig - Director

Overview

The Marine Mammal Research Program is a joint venture between the Department of Wildlife and Fisheries Sciences and the Department of Marine Biology at Texas A&M University, and provides opportunities for students preparing for a career in marine mammal research. The MMRP at Texas A&M is an established and highly respected program, and has worked with and been supported by such agencies as the National Marine Fisheries Service, the Sea Grant Program, and the Marine Mammal Commission. The Program represents a collaborative effort by some of the industry's most capable and accomplished researchers. Its director, Dr. Bernd Wursig, is probably one of the most respected cetacean researchers in the world.

Research Projects

The majority of the Program's research takes place in the Gulf of Mexico, though studies have also been carried out in Costa Rica, New Zealand, and Peru. MMRP researchers cover a huge range of topics, including cetacean foraging strategies, populations, genetics, distribution and residence patterns, and interactions with humans. The MMRP maintains an extensive photographic catalogue for bottlenose dolphins off the coast of Texas, and has also catalogued a large number of sperm whales and Western stock gray whales. MMRP researchers are currently working on a method of identifying dorsal fin photographs through computer software, which could be a tremendous boon to cetacean researchers around the world. The MMRP at Texas A&M is one of the most active cetacean research groups on the planet. It is one of the best places for enterprising marine mammal scientists to turn to for experience and education.

Involvement Opportunities

The MMRP does not make any official use of volunteers, but does provide internship opportunities for college students and recent graduates from all around the world, as well as for graduate students at Texas A&M. Internships last at least three months, and are offered in the spring, summer and fall. Application deadlines are the end of September, the end of February, and the end of May, respectively. Interns work 40 hours per week, and are not paid. Applicants must send a curriculum vitae, a letter of intent, and two reference letters, preferably from professors. The letter of intent should clearly state the term for which the applicant is applying.

Interns participate in field work, which consists primarily of photo-identification and behavioral data collection. The data comes from three or four days of ship-based observations, and is done once every three months. On land, the participants are involved in film processing, Data entry and analysis, and Library research. Interns also participate in Seminars by reviewing and critiquing scientific papers of recent studies. The MMRP usually allows its interns to get involved to varying degrees in a number of its ongoing projects, making a MMRP internship particularly valuable. Housing is not provided, but the Program helps interns find affordable housing. Most participants end up paying $350/month or less, often by sharing apartments with others.

Many researchers in the Program take on a number of graduate students from Texas A&M. Between five and ten active graduate students participate at any one time. Almost all of these are enrolled in MS or PhD programs within the Department of Wildlife and Fisheries Sciences.

Mingan Island Cetacean Study

Nov-May:
285 Green St.
St. Lambert, QC, Canada J4P 1T3
Phone: (514) 948-3669
Fax: (514) 948-1131
Jun-Oct:
124 Bord de la Mer
Longue-Pointe-de-Mingan, QC, Canada G0G 1V0
Phone: (418) 949-2845
http://whale.wheelock.edu/whalenet-stuff/mics.html
rsblues@polysoft.com

If you want fast, heavy-duty exposure to cetacean field research, and you've got a few shekels to spare, you just might want to hook up with the Mingan Island Cetacean Study. For the past twenty years MICS and its founder, Richard Sears, have been running around the waters of the Gulf of St. Lawrence conducting studies on native cetaceans. The organization provides a steady stream of data and results to the scientific community and is responsible for the North Atlantic blue whale photo ID catalogue, but of particular interest to you, O reader, are MICS's extensive and popular research participation sessions.

Research Projects

MICS offers anyone interested in wild cetacean research an opportunity to participate in their research projects, for a reasonable fee. The majority of MICS' research is carried out off the shore of Anticosti Island. The archipelago is a wildlife sanctuary along the north shore of the Gulf of St. Lawrence in Quebec, Canada (about 400 miles north of Bangor, Maine.) MICS has also conducted studies in other parts of the world, including Bermuda, New Caledonia, Zamami (I'd never heard of it either) Iceland, and the Sea of Cortez. In fact, MICS has conducted research off the Baja Peninsula for the past fifteen years, and offers opportunities to assist there as well. Though hard-core field researchers will hmpf at me to say so, the Baja assignment might be little more enjoyable for all the sun and warmer weather. The Baja project focuses on blue whales as well, so if you're after big game, this might be the ticket. Blue whales are also the subject of the group's Icelandic research, and additional work will be done in the Azores islands in 1998. The

Mingan Island sessions run from mid-June through the end of September, while Baja research is carried out during the last week of February and all of March.

Involvement Opportunities

Participants come from many countries, and are not required to speak French, though it would help to be part duck. Much of the research takes place onboard 24-foot rigid-hulled inflatable boats which provide little protection from wind, seaspray or rain. The sessions run from June through September, and last one week apiece. The majority of the week is spent on the water, assisting the researchers with population studies, biopsies, tracking devices, and other projects. On evenings and bad weather days the staff runs seminars and workshops, introducing the guests to biopsy techniques, genetics, marine mammal classification, acoustics, feeding and migration patterns, evolution and adaptations, diving physiology, toxicology, social behavior, anatomy, and whaling history.

MICS offers a number of pricing plans for its sessions. The "Independent Traveler Plan" is for people who wish to secure their own lodgings and food, and simply entitles you to participate in the activities and excursions. It usually costs about $840 Canadian (roughly $700 US.) The "American Plan" hooks you up in a snazzy village motel for all seven nights, transportation to and from the airport and the docks, and breakfasts. In 1997 this cost $1275 Canadian (just over $1000 US) with a bit of a discount for double occupancy. The "European Plan" provides lunches and dinners (though you won't have much of a selection) for $255 Canadian extra. If you agree to participate in both of the June sessions or for all four sessions in September, MICS will give you 10% off any of the pricing plans. (Brr.) The "Student Plan" gets you a plot in a nearby campground and transportation, but no camping equipment or food. (Restaurants and stores are within walking distance.) This option costs just under $800 US. It takes a 50% deposit to hold a spot, and they start filling up rapidly, so call or write early.

Moss Landing Marine Laboratories

Ornithology and Mammology Lab
PO Box 450
Moss Landing CA 95039-0450
Phone: (408) 633-7261
Fax: (408) 633-0805
http://color.mlml.calstate.edu:80/www/mlml.htm
Jim Harvey - Associate Professor
harvey@mlml.calstate.edu

Overview & Research Projects

Moss Landing Marine Laboratories sits on the West Coast, appropriately enough, in Moss Landing, CA. Technically a part of CSU, MLML operates nearly autonomously, and provides research opportunities for interns as well as students. MLML is still rebuilding after the devastation of the 1989 Loma Prieta earthquake, and its facilities are split between Moss Landing and the city of Salinas, ten miles inland. The Lab began construction of new facilities in the fall of 1997, which should be finished by fall, 1999.

A small and relatively quiet community, Moss Landing is inhabited by about five hundred people, most of whom are involved in the local fishing industry. The town is is an ideal location for marine studies. Monterey Submarine Canyon, the largest such feature on the west coast of the Americas, opens just offshore. Nearby is Elkhorn Slough, one of the largest relatively unspoiled salt marshes remaining on the Pacific coast. In combination with the sand dunes, rocky intertidal environments, and subtidal kelp forests in the area, these features give MLML's scientists plenty of plants, biomes, and critters to study.

MLML is operated jointly by seven California State University campuses. Although San Jose State University serves as the Lab's administering institution, graduate and undergraduate classes are open to qualified students of any of the seven campuses. The Lab offers a Masters of Science degree in Marine Science, for which graduate students in the various schools must become classified. (This consists of obtaining a MLML advisor and satisfying both MLML and home campus requirements for coursework and performance.) MLML, by offering the course "MS 112 - Marine Birds and Mammals," is one of the few institutions in the nation to offer an undergraduate class focusing on marine mammals. This is in most part due to the presence of Associate Professor James Harvey, whose background in Marine Mammology is extensive to say the least, and who currently heads up the Laboratories' Ornithology and Mammology Lab.

MLML has assembled a fairly extensive marine sciences library, including over 8,000 monographs, 2,500 bound journals, 92 current subscriptions, a special collection of maps and charts, an extensive reprint file, Aquatic Sciences and Fisheries Abstracts, and a collection of MLML theses and research papers. It's even rumored to have a copy of Moby Dick somewhere.

Involvement Opportunities

Enrolling in MLML's undergraduate classes isn't terribly difficult, but becoming fully classified as a MLML graduate student is, particularly if you wish to focus on marine birds and mammals. Most students who succeed have a 3.2 GPA, scored in the upper 10 to 25% of the nation on their GRE, and have field and/or lab experience in marine vertebrate ecology. (Volunteer or otherwise.) The program supports roughly two dozen students at a time, and competition for classification is intense. The subject matter of recent students' theses has included humpback whales, harbor seals, and harbor porpoises, among other marinelife. Students might employ radio telemetry, molecular analysis, beach surveys, and shore or ship-based observations in their studies, and may even become involved in the local stranding network, in which MLML is very active. Although the marine bird and mammal lab doesn't utilize research assistants at the moment, potential grant-funded research may change that in the future.

Summer internships are also available at MLML, on a volunteer basis. Applicants for Prof. Harvey's program should apply well before February to be considered for the following summer. Applications for internships outside of the summer may be considered, and if so, would probably face less competition. Although MLML provides no financial assistance to interns, they are happy to assist them in finding both employment as well as housing (which tends to be a bit expensive.) There are usually two interns in the mammal/bird program at one time, and they usually have the opportunity to become involved in nearly any aspect of the kinds of studies mentioned above, and might even arrange to take on a project of their own.

Applicants for any of the Ornithology and Mammology programs should contact Jim Harvey directly.

Mote Marine Laboratory

1600 Thompson Parkway
Sarasota, FL 34236
Phone: (941) 388-2451
Fax: (941) 388-4312
http://www.mote.org

Overview

It's hard to classify Sarasota, Florida's Mote Marine Laboratory. On the one hand, it's a tremendous center for marine research, being equipped with state-of-the-art laboratory and computer equipment, and staffed by some of the country's most proven and capable marine researchers. Then again, the presence of the Goldstein Marine Mammal Research and Rehabilitation Center, coupled with the fact that Mote handles the majority of cetacean strandings in southwest Florida, makes a good argument for placing MML in the rescue/rehab category. It also fits many criteria for being a display facility, since its near constant presence of viewable animals and the public Mote Marine Aquarium provide a constant draw for Florida tourists. As if schizophrenia wasn't clearly diagnosed already, MML also offers a limited number of college-credit courses, cooperative research programs with various universities, and an enormous, if unpaid, internship program for college and high school students.

The Lab was born in 1955, making it one of America's oldest marine labs. Tucked away on City Island amid some of Sarasota, Florida's most expensive real estate, the Lab is actually a fair-sized complex made up of two administrative buildings, the aquarium, and the Marine Mammal Research and Rehabilitation Center. The Lab also maintains a remote marine research center on Pigeon Key, allowing researchers immediate access to seagrass communities and coral reefs.

Research Projects

Although MML conducts research into a wide variety of marine subjects, it is particularly well-known for its shark research. I was hoping to see Quint from "Jaws" hanging teeth on the wall, but the staff member on duty informed me it wasn't that sort of establishment. Nevertheless, I was impressed by the sheer volume of information the Lab had amassed on these toothy fish, and I have since learned MML is a world leader in the field. The Lab also carries out a fair amount of cetacean-related research, most of which concerns the local population of Atlantic bottlenose dolphins. Dr. Randy Wells, whose landmark, ongoing study of this group has all but defined the concept of "cetacean population study," is co-sponsored by MML, in conjunction with the Chicago Zoological Society. A full listing of the Lab's research accomplishments would overwhelm the space allotted here, but would include chemical pollution, habitat use and conservation, immunology, environmental chemistry, microbiology, phytoplankton, aquaculture, and coral reef ecology, among other topics. MML welcomes proposals from researchers wishing to utilize their facilities, and accommodates several visiting scientists each year.

Involvement Opportunities

Volunteers and interns supplement more than 70 staff scientists and technicians. Interns come from all over the U.S. and abroad. They are immersed in a wealth of practical experience both in the field and in the laboratory. College interns must be currently enrolled in, or recently graduated from, a graduate or undergraduate program of study related to work being done at MML. For most interns this means biology, marine studies, zoology, or the like, but a number of students in other disciplines have developed atypical internships, such as an Art major working on scientific writing and illustration, or an engineering student developing a new instrumentation buoy. It is not necessary for the school to grant credit for the internship, though some will do so. Interns may have an opportunity to get involved with any animal rehabilitation efforts underway during their visit, should they show an interest.

Volunteers are used at MML for many tasks in almost every aspect of the facility's operation. Guides, cashiers, lab techs, groundskeepers, boat mechanics, just about anybody you meet at the Lab is likely to be a volunteer. Like interns, volunteers are able to assist in the rehabilitation of stranded animals.

Other Programs

Annual memberships for MML directly support the facility's research programs, and entitle the member to free admission, tours, admission to the Monday Night at Mote Lecture (covering a wealth of marine topics), a quarterly newsletter, and discounts on gift shop merchandise. Memberships for students cost $20, and general memberships start at $50.

Nag's Head Dolphin Watch

7517 S. Virginia Dare Trail
Nag's Head, NC
Phone: (919) 441-4112
Rich Mallon-Day - Primary Researcher

One of the newest dolphin studies in the country, the Nag's Head Dolphin Watch is an attempt to map out the bottlenose dolphin groups which frequent the inside waters of North Carolina's Outer Banks. The study is currently funded by a local dolphin-watch company. Over the next few years the current crew will be replaced with students hired to work on the company's tour boat and carry out data collection and analysis. The Nag's Head study will also begin to coordinate the annual ADRC dolphin count for the northern Outer Banks. The project's leader, Rich Mallon-Day, is guardian of the ADRC's website, which will be a good place to watch for updates about this new program. (See page 152 for more information on ADRC.)

New England Aquarium

Harold E. Edgerton Research Lab
Central Wharf
Boston, MA 02110
Phone: (617) 973-5200
Fax: (617) 723-6207
John Prescott - Director

Overview

The New England Aquarium, one of the oldest and best known in the country, has a long history of marine education and conservation efforts. It continues to add to its offerings each year. Work is currently underway on a major expansion which will triple its space and catapult the aquarium into the Information Age with HDTV, virtual reality, and interactive computer exhibitry.

Research Projects

The aquarium maintains a healthy research program through its Edgerton Research Laboratory. The Lab provides valuable information on a vast array of marine subjects to the scientific community and wildlife professionals through peer-reviewed publications, as well as to the public through literature and exhibitry. Studies examine aquatic and marine animals, ecology, conservation biology, comparative biology, and resource management, among other issues.

NEAq research is done in collaboration with such environmental heavy-hitters as the National Marine Fisheries Service, the National Institutes of Health, the Environmental Protection Agency, and the National Science Foundation, and has affiliations with numerous other marine labs both in the U.S. and abroad. Through its research programs the Lab has gained international recognition as an excellent aquatic conservation research center.

Of particular interest to you, O readers, is the Aquarium's right whale program, which occasionally makes use of volunteers. The North Atlantic right whale is the single most endangered whale in the world, having dropped in number to fewer than 350. They were nearly wiped out by the overzealous harvesting techniques of industrial revolution-era whalers. Though even whaling nations recognized the need to extend protection to this nearly extinct species in the 1930's, the whales' recovery has been dangerously slow. This is at least partially attributable to our understanding of the whales' environmental requirements, which until recently has been fairly poor. Habitat destruction, shipping, overfishing and pollution threaten to thwart even the most valiant efforts at rebuilding their once healthy populations. .

The Aquarium is working hard to accomplish exactly that, and has undertaken one of the most extensive behavioral and quantitative analyses of right whales to date. Through photo identification and analysis of population and behavioral data, NEAq scientists are revealing crucial information about the biological and social needs of right whales, which in turn is enabling wildlife management teams to formulate effective plans for right whale recovery.

Involvement Opportunities

Though difficult to obtain, volunteer positions do exist within the right whale program, and offer participants a comprehensive and sometimes harsh look into the reality of conservation biology. Volunteers work closely with the research staff to photograph whales and record behavioral data. They must commit to working for two to three months at a time at remote locations in Maine or Florida, depending on the season. The internships are unpaid, and participants must contribute for their living expenses.

The research staff has become more selective in choosing volunteers in a recent climate of overabundant applicants. Potential volunteers should have extensive experience on watercraft, and some proficiency with cameras. Some ability in quantitative analysis and computer use is also helpful, as are the usual traits of persistence, dedication, and the ability to work with a team. Interested parties should contact the program's coordinator, Scott Kraus.

North Carolina Maritime Museum

Cape Lookout Studies Program
315 Front St.
Beaufort, NC 28516
Phone: (919) 728-7317
Fax: (919) 728-2108
http://www.agr.state.nc.us/maritime
Keith Rittmaster - Dolphin Study Coordinator
kritt@coastalnet.com

Overview

The North Carolina Maritime Museum's history dates back to the turn of the century. The museum is responsible for the preservation of North Carolina's maritime and natural history. The Museum maintains an extremely active education department which offers a huge number of public programs each year. Topics include underwater archaeology, dolphin and turtle behavior, and aquaculture. In addition to classroom and seminar-style programs, the Museum offers and sponsors active participation events like canoe trips, trawl and dredge trips, and overnight naturalist hikes. The Museum's Cape Lookout Studies Program offers groups a chance to custom-tailor their own outdoor educational experience at the Museum's field station on Cape Lookout National Seashore, ten miles southeast of Beaufort. The Seashore is accessible only by boat, and offers a gorgeous view of the surrounding coastline.

Research

One offshoot of the Cape Lookout Studies Program is a population study of local bottlenose dolphins. Initiated in 1985 by Keith Rittmaster and his wife, Vicky Thayer, the study is attempting to map out the locol population of bottlenose dolphins through photo ID work. Rittmaster and his collaborator, Nan Bowles, have constructed a dorsal fin catalogue for these animals, and are seeking both to increase its comprehensiveness and accuracy as well as to put the information to good use in conjunction with other conservation biologists up and down the coastline.

Involvement Opportunities

The dolphin research project relies on a small cadre of individuals who have approached the Museum and requested the opportunity to help out. Data collection trips take place about twice a week. Some volunteers help out on both trips, while others give one day per week. Help is particularly useful for boat maintenance and repair, photograph matching and darkroom work. Volunteers are asked to commit to at least one day per week for three months. Positions are not always available, but the Program is always happy to hear from "dedicated, talented, and energetic" people seeking to help. Interested parties should have their own transportation, an ability to work well with others, and should never run with scissors.

Ocean Mammal Institute

P.O. Box 14422
Reading, PA 19612
Phone: (800) 226-8216
Fax: (610) 670-7386
http://nko.mhpcc.edu/omi/omi_index.html
golub@lclark.edu

Overview

The Ocean Mammal Institute offers the public a chance to play researcher for a little while, exposing them to field research techniques. OMI offers a vast array of cetacean programs ranging from seminars and classes to full-blown, sea-faring data collection trips. Founded by Dr. Marsha L Green ten years ago, OMI has been very successful in documenting the effects of powerboats on local cetaceans in Hawaii. In a field where practical applications of research findings are rare, Dr. Green has had the pleasure of seeing her work support proposed legislation which now helps protect the animals she studies.

Research Projects

OMI's programs are divided into categories targeted at different audiences. Research Expeditions get participants active in actual field research, gathering data, making observations, and recording results. Programs may be land-based, carried out aboard a ship, or a combination of the two. They provide excellent exposure for enterprising scientific-types wishing to get a taste of cetacean ethology. "Ecoadventures" and "Ecoexpeditions" are more suited for the passive participant, giving them a chance to travel, relax, and marvel at the wonders of the sea. "Deep Ecology Workshops" are designed with the spiritual in mind, and several are set aside for women. Of all the research groups covered in this book, OMI seems to represent the most complete fusion between pure science and spiritual pursuits, offering participants a chance to experience one, the other, or both.

Involvement Opportunities

There are no internships at OMI in the usual sense. OMI's internships are scientific expeditions offered to the general public, and incur the same fees as a "Research Expedition". There are opportunities for volunteers to get involved, however. At the time of this writing all of OMI's volunteers were local Hawaiians who expressed an interest in helping out. Although this volunteer "program" is an informal one, OMI shows all the signs of a rapidly growing organization, and more opportunities may open up in the years to come.

Other Programs

OMI offers college courses for graduate and undergraduate students focusing on field research and protecting endangered species, respectively. The credit is arranged through Albright College in Pennsylvania, where Dr. Green is a professor. The undergraduate course takes place in Maui, while the graduate program may take place elsewhere as well.

A naturalist program is offered for individuals who interact with the public on a professional basis. The program provides local educators and guides with the tools they need to captivate and teach others about the marinelife and ecosystems of Hawaii.

Finally, OMI scientists and naturalists give seminars and presentations on a wide variety of environmental subjects.

Oceanic Society Expeditions

Ft.Mason Center Building E
San Francisco, CA 94123
Phone: (800) 326-7491
Fax: (415) 474-3395

Overview

OSE offers two types of expeditions to the public, Natural History and Research. The former is designed for the benefit of the participants, and are guided by an Oceanic Society naturalist and local guides, when possible. The research expeditions are designed to accomplish a specific research goal. Participants work with research biologists as team members, collecting data and logging it into records or computers. Although the research trips actually put the participants to work, they usually allow for some "fun" time as well. OSE offers adopt-a-dolphin and adopt-a-whale programs.

OSE is a reputable, professional organization with a long track record of exciting and productive expeditions. Their staff naturalists are affiliated with such groups as the Nature Conservancy, the National Park Service, U.S. Fish & Wildlife, the Monterey Bay Aquarium, Moss Landing Marine Lab, and a host of Universities and other environmental groups.

Research Projects

The scope of OSE staff experience is impressive, as is the list of expeditions they offer. Recent cetacean-related natural history trips include: orcas at Vancouver Island, humpbacks in Alaska or the Dominican Republic, and gray whales in Baja, Mexico. Some cetacean research expeditions investigated behavioral ecology of dolphins in Belize, ecological needs of Amazon river dolphins, and the social organization and habitat use of spinner dolphins at Midway atoll.

Involvement Opportunities

Embarking on an OSE trip is a relatively painless process. Once someone signs up for an expedition, OSE dispatches information on trip itinerary, entry/exit requirements, a reading list, and insurance information. Three to four months prior to departure, travelers receive flight information, an equipment and clothing list, and educational materials. One month prior, a final detail letter, plane tickets, and a participant list are sent out. Groups are kept small (6-

15 members each) to maximize trip quality and minimize impact on local communities. OSE handles all aspects of the itinerary, and in the case of research expeditions, obtains access or permits which are generally unavailable to the public, depending on the nature of the research.

Most trips include airfare from a select US port city and range roughly from one to three and a half thousand dollars. Write or call or information on current and upcoming programs.

ORCALAB

PO Box 258
Alert Bay, BC
CANADA V0N 1A0
Fax: (250) 974-5513
orcalab@north.island.net
Dr. Paul Spong - Director

If you haven't already surmised from my descriptions of Vancouver and Tacoma, the Pacific Northwest is home to some of the most beautiful territory in North America. ORCALAB rests snugly within a forest along the shoreline of Hanson Island. One of many small islands within Johnstone Strait, Hanson separates the enormous Vancouver Island from the mainland. The strait is a haven to the Northern Resident Population of orcas, which consists of over 190 whales, making this an ideal place to study killer whales.

Overview & Research Projects

ORCALAB is a difficult organization for me to write about. Part of the trouble has been the difficulty in establishing contact with the lab's founder and director, Dr. Paul Spong, whom I suspect is wary of my background with captive cetaceans. The real trouble, however, is in trying capture a unique quality I sense in the scientists and volunteers working with the orcas of the Pacific northwest. Everyone, or at least everyone I have known, is affected in some emotional way by wild cetaceans, a point I have explored in greater detail in the beginning of this book.

The researchers who come to know the orcas seem to feel an even greater connection with these animals than many of us have with other groups of cetaceans. I have little doubt this is due in part to the complex and lasting family bonds exhibited by killer whales. Large populations of orcas are subdivided into cohesive pods and sub-pods, usually organized by order of blood relationships. People participating in long-term studies of these animals are able to chart the progress of mothers and calves, brothers and sisters, cousins and aunts across multiple generations. Although some species of cetaceans exhibit highly complex social groups, none studied have proven as cohesive as the orcas.

Dr. Spong is one of the three North American scientists most knowledgeable in orca behavior and ecology (the others being John

Ford of the Vancouver Aquarium and Ken Balcomb of the Center For Whale Research.) He is in large part responsible for our understanding of orca relationships, having studied and documented them for more than twenty years. In particular, Dr. Spong and ORCALAB made tremendous progress in the recording and analysis of orca vocalizations, and ORCALAB researchers are now able to identify various pods by sounds they make. ORCALAB, and its affiliate non-profit organization, The Pacific Orca Society, continue to investigate the social structure and vocalizations of orcas from their idyllic location on Hansen Island. A network of advanced hydrophones, recorders, and analysis equipment is in the works, as well as a submerged observation complex. As impressive as orca research has been so far, it's a good bet much remains to be discovered beneath the waters of Johnstone Strait.

Involvement Opportunities

There are no advertised positions at ORCALAB, though the facility has taken on interns in the past. These internships lasted six weeks or more, and have generally involved photo identification of whales, recording and analysis of vocalizations, and the tracking of orca movement patterns.

Pacific Cetacean Group

UC MBEST Center
3239 Imjin Rd.
Marina, CA 93933
Phone: (408) 582-1030
Fax: (408) 582-1031
P007CG@aol.com
http://infomanage.com/pcg

Overview

The Pacific Cetacean Group is a nonprofit organization dedicated to "increasing the knowledge of marine mammals and promoting the conservation of their marine habitats through scientific research and education." Headquartered in the Monterey Bay area, many of its programs focus on the wildlife and ecology of that area. Possibly the most relevant tidbit about the PCG (insofar as this book is concerned) is that the marine scientists and educators who run the organization want to get get the public involved. They offer numerous research and education programs which focus specifically on marine mammals and their environment. PCG's office and classrooms are situated in the University of California, Monterey Bay Education, Science and Technology Center, at Fort Ord in Marina.

Research Projects

Much of PCG's research focuses on monitoring populations of cetaceans and other marine mammals by collecting long-term data on the abundance, distribution, behavior, and feeding and breeding ecology of bottlenose dolphins, humpback and blue whales, and sea otters.

The PCG assists other agencies in monitoring human induced disturbances and naturally caused phenomena which affect marine habitats, such as pollution, El Nino, and red tides. PCG staff have worked with research teams at Orange Coast College and San Diego State University in a long-term study of bottlenose dolphins along the entire California coast.

PCG is developing a long-term monitoring study of the ecology of marine birds and mammals in Monterey Bay in which students will participate in the collection and analysis of survey data.

Involvement Opportunities

The Group runs educational boat trips, out-reach programs, presentations, and research internships for high school and college students. PCG offers the public opportunities to participate in its research programs for a reasonable fee. Contact the Marina office for more information.

Other Programs

A membership in the PCG gets you a tri-annual newsletter, updates on upcoming events, a membership card, and a 10% discount on their research and education programs, whale watching trips, and coffee mugs (and other merchandise). Many levels of memberships are available; the regular rate is $25, $20 for students, and $10 to receive just the newsletters. Checks should be payable to The Pacific Cetacean Group, and sent to their main address above.

Pacific Whale Foundation

Kealia Beach Plaza
101 N. Kihei Rd., Suite 21
Kihei, Maui, HI 96753
Phone: (808) 879-8860
Fax: (808) 879-2615
pacwhale@igc.apc.org

Overview

The Pacific Whale Foundation is a nonprofit conservation group whose aim is to collect statistical and behavioral data on cetacean populations of in the Pacific ocean. Located in beautiful, scenic Maui, the Foundation is dedicated to gathering and updating scientific data used for whale conservation.

Research Projects

At the moment the PWF focuses on Humpback whales along the Hawaiian, Japanese, Alaskan, and Australian coasts, as well as spinner dolphins and other reef residents in Hawaii. In both cases researchers collect data through non-invasive techniques, mostly photo I.D. and shore-based observations. They publish this information in scientific journals and hand it over to resource management agencies. This allows for informed decisions concerning the species' survival and well-being. PWF research has been instrumental in the formation of marine sanctuaries in the United States, Australia, and New Zealand.

Involvement Opportunities

There is no formal volunteer program for their projects, but they do offer field internships in Hawaii and Australia, for a fee. The cost, which covers all expenses except airfare to the site, usually ranges from $1500 to $1900, though discounts of $500 have been offered in the past for early enrollment (January - February). They usually require an advance deposit of $250, $200 of which is refundable up to 45 days prior to the trip. The teams are limited to four or six people, with only eight teams per location, so it's definitely a good idea to book well in advance.

Other Programs

The Foundation also runs a number of educational programs, including a traveling classroom van, an adopt-a-whale program, reef walks for children, guided whale watches, and a marine mammal naturalist certification course. Memberships are also available, which entitle you to the Foundations newsletters and periodic notifications on environmental topics. Individual memberships are $30, $25 for students, and can be paid by check (payable to Pacific Whale Foundation) or by credit card (Visa, MC, and Amex). Send questions or requests to the above address.

Project Pod

718 Fisherman Wharf
Fort Myers Beach, FL 33931
Phone: (941) 765-8101
Fax: (941) 765-5458
http://swflorida.com/dolphin
dolphman@peganet.com
Joël Bellucci - Project Director

Overview

Project Pod, the brainchild of Florida environmentalist and 3D computer artist Joël Bellucci, is an ongoing investigation of the bottlenose dolphins inhabiting Estero Bay, the body of water separating Ft. Myers Beach from the mainland. Created in 1995, the Project was originally sponsored in part by the Ostego Bay Foundation, a local marine education and conservation group. The Project soon attracted the attention of the Frederick S. Upton Foundation, which provided an operational budget for the next two years. The Project is being adopted by the Estero Bay Marine Lab, which is a research offshoot of the Ostego Bay Marine Foundation and the newest of America's marine laboratories.

The first two years of the Project have been spent compiling a photographic directory of the bay's dolphins. Still in its infancy, the second phase of Project Pod will focus on behavioral data, and involves weekly trips through the bay aboard the Project's 21-foot skiff, taking photographs, video footage, and behavioral notes. Project Pod is an excellent example of the opportunities which abound for enterprising researchers to examine and document resident cetacean populations along the North American Coastline.

Employment, Intern, and Volunteer Opportunities

Project Pod has no paid positions, but the Ostego Bay Foundation takes on unpaid interns who participate in the Project's data collecting trips. OBF interns get involved in more than a dozen other research projects, covering local tidal flow, plankton levels, water quality, and other marine issues. Interns help with the Foundation's considerable efforts to reach out to local communities through educational programs. Internships typically last from four to nine months, depending on how much time the intern wished to give. Candidates need not be currently enrolled in a college or university, but must provide their own lodgings and transportation.

Interested parties should send a resume and letter to:

Ostego Bay Foundation
Joanne Semmer - President
718 Fisherman's Wharf
Ft. Myers, FL 33931
Phone: (941) 765-8101
Fax: (941) 463-0865

Project Pod has made use of local resident volunteers for its excursions and data analysis, but the Project cannot guarantee participation on a weekly basis.

Virginia Marine Science Museum

717 General Booth Boulevard
Virginia Beach, VA 23451
Phone: (757) 437-4949
Fax: (757) 437-6363
http://www.whro.org/vmsm
Mark Swingle - Primary Researcher

Overview

VMSM is a certified Full-Fledged Public Treasure House of displays and programs describing the habitats and wildlife along Virginia's coastlines. It features over 150 exhibits, an Imax screen, and an entire salt marsh. River otters, sea turtles, and a plethora of fish and birds are seen by thousands of visitors each year, making the Museum one of Virginia's most popular tourist attractions. The Museum also features interactive exhibits on marine mammals and scale models of Virginia's most common cetaceans in its new Ocean Pavilion.

Research Projects

The Virginia Marine Science Museum conducts a fair amount of marine research in addition to providing Virginia with one of the country's best sources of information and exhibitry on marine subjects. In particular, the Museum's staff has been carrying out a photo identification study of indigenous dolphins and whales, particularly bottlenose dolphins.

Involvement Opportunities

While the dolphin study at VMSM does make use of a handful of the Museum's volunteers, the staff is fairly selective, and it can be a difficult program to get into. The most successful candidates will be local residents who build a relationship with the staff by volunteering in other areas for a time. Volunteers need only be 18 years of age, and willing to commit to a regular weekly shift.

Other Programs

VMSM is an active member of the Marine Mammal Stranding Network, and maintains a list of volunteers to assist with response efforts. The Museum maintains two teams of volunteers to assist with stranding: an actual response team, and a core of administrative and support volunteers. Most of the stranded cetaceans the

Museum encounters are dead. When a live animal is found, the Museum staff may decide to utilize its modest holding facilities to rehabilitate it, at which point volunteers from both teams may be asked to assist with the animal's care. This procedure has not been done with a cetacean in recent years, and such animals would most likely be brought to the National Aquarium in Baltimore, if possible.

Stranding volunteers must be 18 years of age, commit to a certain number of hours each month, and must live near the Virginia coastline. Stranding volunteers assist with live animal rescue, rehabilitation, carcass recovery, and necropsies. VMSM also offers unpaid internships to college and graduate students. Interested students are encouraged to contact the VMSM volunteer coordinator and ask about internship opportunities within the Stranding Program.

Whale Conservation Institute

191 Weston Road
Lincoln, MA 01773
Phone: (781) 259-0423
Fax: (781) 259-0288
Dr. Roger Payne - President
kim@whale.org

Overview

Easily a candidate for either the Conservation or Research sections, the Whale Conservation Institute is an exclusively scientific and educational organization which has committed itself to "protecting and conserving whales through ground-breaking research and international education initiatives." In particular, it has been compiling and distributing data on humpback and right whales since 1971. It was created by Dr. Roger Payne, one of the United States' foremost cetacean researchers. In addition to conducting over one hundred studies involving every species of large whale over the past three decades, he is best known to the general public for co-discovering that the long and complex vocalizations of humpback whales are songs.

Dr. Payne is still president of WCI, which supports his ongoing studies of southern right whales in Argentina and humpback whale songs. The photo-identification work of the right whale program has amassed data on more than 1200 individuals over the last 28 years. (This is, in fact the longest running study of any great whale). WCI is particularly proud of their development over the years of benign, non-invasive methods of collecting data, many of which have been adopted by other researchers around the world.

Recognizing the almost uncontrollably deteriorating situation faced by whale populations today, WCI has been knocking itself out trying to show the public how serious things are, and what can be done about them. Outreach programs have brought WCI conservationists and scientists into schools across the nation armed with slides, interactive video presentations, and "Mike", the inflatable walk-through fin whale. Educational curricula for use in grade, middle, and high schools have been developed by WCI with an emphasis of environmental awareness. A partnership with a local whale watch company, the Cape Ann Whale Watch, has given WCI naturalists an additional 50,000 interested listeners each year to educate. As if that weren't sufficient, movies, television productions, books, interactive laserdisks, an Omnimax film, and an interactive internet project allowing students to track tagged whales have all added to WCI's global efforts to teach environmental awareness and foster responsible behavior.

Research Projects

In addition to Dr. Payne's continuing research on humpback whales and southern right whales, WCI conducts oceanic research aboard a 93-foot ketch donated by a Chicago businessman and, of all people, the maker of the board game Trivial Pursuit (a perhaps less than fortunate term for a whale research vessel to be associated with.) The Odyssey is a floating research platform from which WCI investigates organohalogen toxin levels, the diving, feeding, and migration behavior of sperm whales, and bioacoustics. WCI also maintains a digital database of photographed whales which it shares with other cetacean researchers worldwide. Marine pollution, DNA fingerprinting, and bioacoustics have become the main focus of WCI's research, and although the results already obtained have been promising, future findings may prove to be even more so.

Involvement Opportunities

At the time of this writing, WCI had only just begun to finalize a new volunteer and internship program. Check the WCI website, or contact them directly for the latest information. Memberships in WCI cost anywhere from fifteen to five hundred dollars per year, depending on the level desired. All include WCI's bi-annual newsletters and most include a whale adoption.

6
Conservation Groups

You've probably heard your Aunt Marge say "This world's going to Hell in a handbasket". That isn't too far from the truth. We're facing evaporating rainforests, our water tastes awful, and entire species of animals are disappearing. Sea Cows, for example, were twenty-six foot long, three thousand-pound manatees from the Bering sea, but you'll never see one, since hunters killed the last of their kind in 1769. Lots of animals are headed for a similar fate, including many species of whales and dolphins.

What is Cetacean Conservation?

Some whale species would be dead right now if it weren't for the efforts of conservation groups. Conservationists identify sources of environmental problems, relay the information to governments, industries, and the public, and fight to bring about beneficial changes. Whaling, one of the most obvious causes of dolphin and whale deaths, is still a hot topic.

Whalers existed hundreds of years ago, but the invention of explosive harpoons and factory ships after the Industrial Revolution threw everything out of balance. Technology suddenly handed whalers a huge advantage, and whales started dying faster than they could reproduce. The 1986 international ban on whaling helped slow down the disappearance of many cetaceans, but the fight continues. Several industries and even some nations keep trying to dismantle the ban. More pressure on vulnerable species comes from habitat destruction and pollution. Officials treat beluga whales from the Gulf of St. Lawrence in Canada as toxic waste when they wash ashore because of the high level of pollutants in the whales' bodies. Despite the best efforts of environmental groups worldwide, the battle may still be lost.

Here are some words from Dr. Roger Payne, the Founder and Director of the Whale Conservation Institute:

> "There's no such thing as a species of large animal whose survival can be ensured. Each new generation of humans creates fresh threats to (their) existence. Despoilers need but win once and a species is lost forever.
>
> "The nuclear age conditioned us to believe that the end of the world would be violent - some super weapon in the hands of a madman, or some natural disaster. One of the peace benefits of the end of the cold war has been the opportunity to reflect on less dramatic but more likely scenarios - to realize that life as we know it is more likely to end 'not with a bang, but with a whimper'.
>
> "What will kill us and the rest of life on earth now seems most likely to be a slow graying of life; a weakening of immune systems, disruptions in normal development, devastating genetic aberrations and mutations. Scenarios that just a few years ago seemed improbable now seem likely - if not to end life outright, then to transform it into something you and I would not wish to be part of, nor have our children endure.
>
> "To avoid this we must act with an intensity we have never before approached. As ocean pollution levels increase, marine mammals like whales will be among the first to go. They are like the canary in the coal mine, warning us of dangers we cannot sense ourselves. Whales alas, are far from being saved. In many ways the job has just begun."

How can I join the fight?

If you feel the call to help defend dolphins and whales from extinction, all you have to do find a group, and jump in. There are opportunities for scientists, activists, and hobbyists alike. Some of the fourteen organizations in this chapter fight whaling, tuna boat bycatch, and culling operations. Others try to eliminate drift nets, chemical pollution, and marine debris.

Most conservation work is done from an office. Expect lots of phone calling, faxing, mailing, and envelope stuffing. Even the direct-action groups like Greenpeace and Sea Shepherd spend long hours gathering support and disseminating information.

None of these groups have very many opportunities for employment. Help out for free as much as you can. If dolphin and whale conservation continues to be the driving force in your life, all you need to do is build up enough experience. Sooner or later the groups you work with will offer you a job. Of course, your chances improve if you build skills they can use. Pro-bono services from lawyers are always in demand, and writers, researchers, stuff scroungers, public speakers, and experienced grant writers can find ways to help out too.

Conservation Organization Descriptions

The descriptions on the following pages are organized just like the Research entries, except "Conservation Efforts" has been substituted for "Research Projects". Other conservation groups exist; you'll only find those whose mission is substantially linked to cetaceans here.

American Cetacean Society

P.O. Box 2639
San Pedro, CA 90731
Phone: (310) 548-6279
Fax: (310) 548-6950

Overview

The main goal of the American Cetacean society is to protect cetaceans and their habitats through public education, research grants, and conservation actions. Founded in 1967, It is the world's oldest cetacean conservation organization. (Try saying THAT ten times really fast!) ACS is a non-profit, volunteer organization with several regional U.S. chapters and members in 41 countries. The ACS's main headquarters is in San Pedro, California.

Conservation Efforts

The ACS uses education as its primary tool in dolphin and whale conservation. By serving as a clearing house for cetacean-related information, ACS manages to educate thousands of people each year through informational brochures, packets, and books as well as written responses to individual queries.

ACS is very active in the area of cetacean research by funding individual researchers (graduate students as well as established biologists) and by providing access to the results of the research they fund through publications and annual conferences. They also maintain a central research library in San Pedro. (Which, incidentally, would be a valuable resource for tracking down the names of cetacean scientists.)

Every two years the ACS holds a national conference - a unique opportunity to gather information almost any type of cetacean-related group and a chance to mingle among some of the world's brightest cetacean researchers and conservationists. The next conference, which will most likely be held near San Pedro, is scheduled for the winter of '98.

Involvement Opportunities

ACS is an excellent source of information, both through its membership publications and its conferences. Occasionally, regional offices use volunteers to help with conservation efforts; such opportunities are announced in literature sent to members. An ACS membership costs only $35 annually ($25 for students,) which is a bargain for the information you'll get from their newsletter, journal, and Whalewatcher, ACS' regular report on cetacean research. Volunteering your time at one of ACS' seven chapters is also a good way to get involved in cetacean conservation, and possibly a good place to find leads for field research. Below is the contact information for ACS's local chapters:

Pacific Northwest
PO Box 22096
Juneau, AK 99802

Los Angeles
(310) 548-8500

San Diego
(619) 286-8717

Santa Barbara
3930 Harrold Ave
Santa Barbara, CA 93110

Orange County
PO Box 18763
Irvine, CA 92623
(714) 534-5177

Monterey Bay
PO Box HE
Pacific Grove, CA 93950

NY/New Jersey
PO Box 392
Center Moriches NY 11934
(718) 945-1677

Center For Marine Conservation

1725 DeSales NW Ste. 600
Washington, DC 20036
Phone: (202) 429-5609
Fax: (202) 872-0619

Overview

The Center for Marine Conservation is the largest of the marine conservation groups, and possesses one of the most dedicated and driven staffs in the conservation community. It was founded in 1972, a time when public concern over marine pollution was on the rise, and the same year as the inception of the Marine Mammal Protection Act. The Center set out to reduce marine pollution, and protect marine habitats, fisheries, and endangered marine animals. It has succeeded in all four areas, and continues to bring marine conservation issues to the attention of the public and the government.

Conservation Efforts

In 1986 The Center helped establish the world's first whale sanctuary in the Dominican Republic. It has been involved in numerous legislative debates which affect cetaceans, including the dolphin-tuna issue, the establishment of the National Marine Sanctuaries, and reauthorization and amendments for the Endangered Species and Marine Mammal Protection Acts. The Center has been instrumental in the preservation of the Clean Water Act, the implementation of oil spill protection laws, and is active in over a dozen political battles over environmental issues ranging from sea turtle protection to marine debris and drift nets.

The Center maintains a number of satellite offices (listed below), which enable the Center to focus on local issues and to deliver its messages to the public more effectively. These regional and field offices may provide a starting point for getting involved in marine conservation for local residents.

Involvement Opportunities

The Center does maintain a sizable paid staff, though openings are rare. There are about 10 employees at the Washington office, and approximately 60 employees nationwide.

The Center encourages its members to become active volunteers and assist in its efforts. The vast majority of the Center's volunteer needs concern administrative tasks, fundraising efforts, and beach cleanups. The Center sponsors a number of specific marine conservation programs which utilize volunteers. "A Million Points of Blight" is a campaign to keep debris and pollutants out of storm drains and which uses volunteers to stencil warnings on drains across the country. The "National Marine Debris Monitoring Program" trains volunteers to record important data on debris found on beaches. Scuba-diving volunteers in the "Underwater Cleanup and Conservation Monitoring Program" help remove undersea trash, and the "Cruise Line Watch" teaches passengers how to monitor the environmental-friendliness of Cruise Lines. Contact the Center's Washington office for information about memberships and volunteer opportunities.

The Center's International Coastal Cleanup is held annually on the third Saturday of September. It is the world's largest volunteer effort dedicated to the marine environment. In 1996 more than 300,000 volunteers collected over four million pieces of debris on land and underwater throughout the United States and some 90 other countries.

Other Programs

To further its goal of educating the public about marine conservation, the CMC offers a variety of information packets, brochures, posters, stickers, and audio-visual materials about related subjects, all of which are available from the main office. Listed below are the numbers and addresses for the Center's regional offices:

Beach Office 1432 N Great Neck Rd. Suite 103 Virginia Beach, VA 23454	Phone: (757) 496-0920 Fax: (757) 496-3207
Regional Office 1 Beach Dr SE #304 St. Petersburg, FL 33701	Phone: (813) 895-2188 Fax: (813) 895-3248
Key West Office	Phone: (305) 295-3370
Pacific Regional Office 580 Market St. Suite 550 San Francisco, CA 94104	Phone: (415) 391-6204 Fax: (415) 956-7441

Cetacean Society International

PO Box 953
Georgetown, CT 06829
Phone: (203) 544-8617
Fax: (203) 544-8617
http://elfi.com/csihome.html
71322.1637@compuserve.com

Overview

Way back in 1974 a group of dolphin and whale enthusiasts set up shop as the Connecticut Cetacean Society, and tried to make a dent in the large and growing wall of serious problems facing dolphins and whales. Twenty-three years, twenty-six countries, and a name-change operation later, the Cetacean Society International now presents a small but influential presence at nearly every conference and debate affecting cetaceans in the wild. They partially fund independent researchers in many parts of the world, and seem especially proud of their progress in bringing modern, effective studies of cetaceans to underdeveloped nations.

Conservation Efforts

CSI disseminates information about whales through whale-watches, special presentations, and responses to individual queries. They are involved with several new educational programs including support for local dolphin and whale conservation efforts in Peru, Chile, Argentina, Brazil, and Columbia. CSI publishes a very informative newsletter, Whales Alive!, both on paper for its members, and also at CSI's web site (see URL above.)

CSI is a non-profit, all-volunteer organization with an impressive amount of experience behind it. In addition to its efforts at wild cetacean conservation, CSI seeks to bring intelligent discussion to the always murky issue of captive dolphins and whales. Although they argue against the capture of new animals, they adopt an open stance concerning the disposition of extant display animals, captive-born, and stranded animals unfit for release. Regardless of its opinions, CSI approaches this and any issue through responsible and well-informed debate, which, I might add, is quite refreshing.

Involvement Opportunities

If you're looking for a place to volunteer your time, CSI provides an excellent opportunity to donate not only your time and effort, but a part of yourself as well. Having no formal volunteer program, CSI encourages potential volunteers to help identify the ways in which their particular skills and abilities might benefit the organization. If you're a database engineer, you could probably dream up some nifty new CSI information network. If you're in marketing, maybe you can help them increase their membership. If you're a redneck, tell them you'll organize the first CSI Annual Whale Benefit Hoe-down. Whatever your forté, being able to bring something personal to a volunteer program is a very rewarding experience.

Membership for U.S. residents is $15/year and includes the Whales Alive! newsletter, membership meetings with cetacean experts from around the world, whale watching trips, and access to CSI's not inconsiderable collection of cetacean books and videos in Georgetown, CT. Other membership levels are available.

Cousteau Society

870 Greenbrier Ste. 402
Chesapeake, VA 23320-2641
Phone: (800) 441-4395
Fax: (757) 523-2747

June 25th, 1997 was a sad day for anyone with a fascination for the sea and the mysteries and wonders beneath its surface. Jacques-Yves Cousteau passed on after decades of pioneering marine research and conservation. He was the embodiment of our fascination with the sea, and will be remembered forever as the father of underwater exploration. To aspire to lead a life even half as full as his is a worthy ambition indeed.

Overview

The Cousteau Society was formed by Captain Cousteau in 1973 as a vehicle for both marine conservation and research. Naturally, the organization was founded in the United States, but it wasn't long before members from other countries began to appear, and the organization now has an international membership of 150,000. The organizations research and expeditions have been the subject of more than forty books, eight filmstrips, four feature films, and over a hundred television documentaries, all of which has contributed to making the Society one of the world's most high-profile marine research and conservation groups.

The Society takes a holistic approach to marine research, paying equal attention to the oceanographic, biological, atmospheric, and cultural aspects of a given region. Past studies have taken place in Haiti, Cuba, the Marquesas Islands, New Zealand, Australia, Papua New Guinea, Thailand, Borneo, Indonesia, Madagascar, and South Africa, among other places.

While the Society has undoubtedly contributed an unprecedented amount of research to the fields of oceanography, marine biology, and marine engineering, it is listed here as a conservation group because it has no accessible ongoing cetacean research projects. One never can tell, though, where the research vessel Alcyone or its soon-to-be sister ship, Calypso II, will sail to next, and The Cousteau Society may yet provide opportunities for enterprising cetacean researchers in the years to come.

Conservation Efforts

The Society has always been committed to preserving our planet for future generations, and has worked hard to ensure stable and unpolluted habitats for all marine life, including cetaceans. It was instrumental in launching the Protocol on Environmental Protection to the Antarctic Treaty, an initiative to protect the world's last great wilderness, and is hoping to succeed in persuading Russia and Japan, the last two members of the treaty to ratify the Protocol, to sign on in the near future. The Society has its hands in a number of other international conservation efforts, and remains as committed as ever to protecting the world's waters.

Involvement Opportunities

Primary support for the Society's efforts comes from the annual dues of its members. Memberships are available in a variety of amounts, beginning at $30 for the basic model. Members receive a subscription to Calypso Log, the Society's bi-monthly periodical. Family memberships also include Dolphin Log, a bi-monthly aimed at younger readers.

Earth Island Institute

300 Broadway, Suite 28
San Francisco, CA 94133-3312
Phone: (415) 788-3666
Fax: (415) 788-7324
http://www.earthisland.org/ei
marinemammal@igc.apc.org

Whales Alive!

PO Box 2058
Kihei, Maui, HI 96753
Phone: (808) 874-6855

Overview

Earth Island Institute is one of the most vocal and active conservation groups in the United States, attacking big, beefy environmental issues like global climate, deforestation in rainforests and temperate old growth forests, marine habitat destruction, and the preservation of endangered species. Earth Island Institute's International Marine Mammal Project is devoted to the elimination of commercial whaling, the use of dolphins and whales in public display facilities and therapy programs, incidental dolphin killings within the tuna industry, and drift nets.

Conservation Efforts

The Institute has been remarkably effective in combating dolphin deaths within the tuna industry. In 1987 it sent an undercover cameraman aboard a tuna boat and obtained telling footage of how many dolphins were caught and killed. Not long after, the group succeeded in obtaining a court decision to include a federal observer aboard each tuna fishing vessel. The group has been equally active in the recent events surrounding NAFTA and the potential weakening of "dolphin-safe tuna" legislation, and the reauthorization of the Marine Mammal Protection Act.

Earth Island Institute also maintains a presence at International Whaling Commission meetings, and frequently lends it voice to the numerous conservation groups which typically lobby for animals' rights at the meetings. One such occurrence was an effort in 1996 to prevent a U.S. proposal to allow subsistence whaling of gray whales by the Makah tribe. Such a decision would have run contrary to existing standards which require a native people to have depended on subsistence whaling continuously in the recent past. (The Makah have not hunted whales in over 70 years.) The proposal was being pushed, unsurprisingly, by the Japanese and Norwegians, and suffered a bit of a setback when a group of Makahs arrived at the conference to oppose the idea. (Doh!)

The Institute has made several efforts to persuade zoos and aquariums to stop the display of dolphins and whales in artificial habitats. It lent its support to fund-raising campaigns in support of Keiko's rehabilitation (the orca star of the Free Willy movies), and has organized numerous letter-writing campaigns to help end the continued capture of cetaceans from the wild. Current efforts are aimed at generating enough of a public response to affect the reauthorization of the Marine Mammal Protection Act in 1998.

The Institute is the parent organization of "Whales Alive!", a whale conservation effort in Hawaii. Whales Alive! offers public whale watching excursions, which it uses to educate the public about whales and the environmental problems they face. The group is also actively defending the newly-formed Hawaiian National Marine Sanctuary, which has run into ratification troubles with the Hawaiian state government.

Involvement Opportunities

Earth Island Institute's primary need for assistance is with its letter-writing campaigns and membership drives. The San Francisco office makes use of volunteer assistance, and occasional opportunities arise within the organization for internships. Whales Alive! also makes occasional use of volunteer assistance in Hawaii.

Earthtrust

25 Kaneohe Bay Drive, Suite 205
Kailua, HI 96734
Phone: (808) 254-2866
Fax: (808) 254-6409
http://planet-hawaii.com/earthtrust
earthtrust@aloha.net

Overview

Sherlock Holmes fans will find Earthtrust a particularly exciting group to hang out with. It is an international nonprofit organization which has maintained a watchful eye on the status of endangered cetaceans, and has taken on an investigative role in the fight against illegal whaling. Earthtrust members have been present at International Whaling Commission meetings since 1979, and the group is providing indispensable information to the international legislative and scientific communities about illegal whaling and fishing activities as well as cetacean behavior and cognition.

Conservation Efforts

Earthtrust scientists developed techniques for testing the DNA of meat suspected to be from illegal whaling operations. This is a critical step in enforcing the international moratorium on whaling, and in catching illegal sales of whalemeat.

Japan is the primary market for illegal whalemeat, where the price for a pound of flesh from an endangered whale can reach hundreds of dollars. Japan circumvents the moratorium with a modicum of legitimacy through the use of "scientific permits." (Any IWC nation can issue such permits to itself, and is not required to justify them to the rest of the IWC.) There is recent evidence to suggest that the meat from these "scientific" catches is being dumped on the market, possibly to explain the otherwise mysterious availability of the stuff. Once on the market, it is nearly impossible to tell whether a sample of meat is from a scientific permit animal (which would have to be a non-endangered species) or if it came from a threatened species taken by a "pirate" whaling vessel. The DNA testing developed by Earthtrust can tell enforcement agencies exactly what kind of animal the meat is from. This is a big step toward eliminating illegal whaling operations!

Interestingly enough, this whole situation is moving into some pretty weird legal waters. The IWC doesn't have any enforcement abilities, and no single nation will send spies into foreign countries over whalemeat. Consequently, it is falling to non-government organizations like Earthtrust to equip themselves with not only the scientific expertise to design the tests, but to gather the discrete talent necessary to carry the tests out unnoticed. This has given rise to the use of environmental secret agents, people who can enter a foreign country with the necessary equipment and track down dubious meat without raising suspicion. (Now look, don't everyone run out and send Earthtrust your resumes to become eco-spies, or I'll hear about it for years.)

Earthtrust pushed for years to end the use of drift nets, and was partially responsible for convincing the U.N. to ban them in 1993. They are now working towards eliminating pirate drift net operations. Pirate fishing operations are notoriously bad at keeping records, so there isn't any solid data on how many there are or how much damage they continue to do. It's probably a fair guess to say far too many and far too much. Earthtrust continues to seek out and document such activity as much as possible.

Earthtrust is involved with research on dolphin intelligence and cognition at Sea Life Park in Hawaii. The research, headed up by Dr. Ken Marten, seeks to determine whether dolphins are self-aware through the innovative use of mirrors and televisions. Some of this study's results may be seen in the paper "Evidence of Self-Awareness in the Bottlenose Dolphin" in the book Self Awareness in Animals and Humans. (See reading list at the end of the book.) They also sponsor the work being done at ORCALAB, and will doubtless continue to expand their research efforts in the future.

Involvement Opportunities

Earthtrust does occasionally take on interns, usually people dedicated to global conservation who are willing to work for a minimum of two months for free. Previous interns assisted with Earthtrust's dolphin cognition program and helped photograph wild dolphins. Most of their intern work (and, they point out, the majority of conservation work today) involves office equipment and supplies. Earthtrust is always in need of highly skilled volunteer assistance, so if you have a skill you feel might be helpful, give them a call. Most internship and volunteer opportunities with Earthtrust are in Hawaii, though they are currently recruiting "Earth Nerds" (computer geeks like myself) from all over.

Great Whales Foundation

PO Box 6847
Malibu, CA 90264
Phone: (415) 458-3262
Fax: (310) 317-1414
http://elfi.com/gwf.html
whales@elfi.com

Overview & Conservation Efforts

The GWF bills itself as "an extremely intense all-volunteer group" which has taken on the task of representing cetaceans' interests both in the public sector as well as within legislative bodies. They are skeptical of cetacean researchers in the employ of the government or "dolphin-using industries", and seek to identify, support, and distribute the information gathered by what they consider to be reputable and unbiased researchers. The GWF does not agree with the concept of managing cetaceans as a sustainable resource, preferring instead to assign the animals a certain degree of intrinsic rights within our society. In addition to distributing information through educational meetings and interactive media, the GWF has been active in supporting pro-cetacean delegates at the International Whaling Commission's annual meetings.

Involvement Opportunities

The Foundation does not offer memberships, but does welcome volunteers who wish to lend their skills in a variety of support roles. Most of their volunteers help with information distribution, particularly through electronic media. Interested parties should contact the headquarters office listed above.

Greenpeace and Sea Shepherd

Greenpeace USA
1436 U St. NW
Washington, DC 20009
Phone: (202) 462-1177
Fax: (202) 462-4507
http://www.greenpeaceusa.org

Sea Shepherd Conservation Society
PO Box 628
Venice, CA 90294
Phone: (310) 301-7325
Fax: (310) 574-3161
http://www.seashepherd.org

Some people may criticize me for discussing these two groups in the same section, but hey, some people shave their heads too, so there you are. There is no real link between Sea Shepherd and Greenpeace now, but their origins are very closely related, and the two present an interesting disparity of methods.

Overview and Conservation Efforts

Both seek to end whaling and bycatch of cetaceans in the fishing industry, but Greenpeace's goals, and its resources, are much larger. The group was formed in 1972 by a band of nuclear test protesters from Vancouver. They were determined to stop the detonation of a five megaton device under the Alaskan island of Amchitka, at the tip of the Aleutian chain. Although the fledgling organization failed to stop the test, the political fallout resulting from their actions pressured the U.S. government to find someplace else to play with its toys, and Amchitka remains a wildlife sanctuary today. Greenpeace continued to protest nuclear testing, and has developed quite a track record for bringing global attention to the issue. (Of particular interest is their long and incredible history with French riot police, commandos, and secret agents. The French government is nothing short of astounding when it comes to making foreign policy decisions. It's a wonder, for example, that New Zealand hasn't declared war on them yet. Check out Greenpeace's autobiography, The Greenpeace Story, for a detailed account.)

After just a couple years of being an anti-nuclear test group, Greenpeace was approached by Dr. Paul Spong (see ORCALAB, page 201) about trying to stop the Soviet and Japanese whaling fleets from operating off the West Coast. He and a couple of other Greenpeace members formed a group called Project Ahab to address the issue, but it wasn't long before the effort was integrated into Greenpeace itself.

It was the first of many new branches in Greenpeace activity, and the organization is now concerned with nearly every environmental issue imaginable. It has become, in the public eye, the world's leader in environmental protection. More than 700 paid Greenpeace staff run offices in 32 countries. Acting as an intermediary between them all is Greenpeace International, based in Amsterdam. Greenpeace International is funded by the national offices, who in turn are financed almost entirely by small contributions from 2.9 million supporters in 158 countries, and by sales of merchandise.

Sea Shepherd has a much, much smaller membership and budget but depending on who you talk to, it may have done more to protect large whales than any other group of a comparable size. Sea Shepherd is a "direct action" group, which means they are less than diplomatic and more than expressive in their objections. Ships are the organization's primary weapon in the fight to protect whales. The Sea Shepherd fleet is crewed entirely by volunteers which patrol the waves in search of illegal whaling operations.

To the best of my knowledge (which is admittedly somewhat patchy) Sea Shepherd adheres to its stated policies, which claim that "every precaution must be taken to ensure that no human life is taken or that any injury will be caused to any human being during an operation." These principles underscore the group's compliance with national laws and international conventions. (Indeed, they draw their authority to enforce whaling laws from just such a document.)

Still, there exists a certain subtext in Sea Shepherd's literature, made evident when they take the time to point out that no member has ever been convicted of any violation, which suggests a willingness to test the boundaries of "acceptable" behavior. Their primary goal, as stated, is to identify and document illegal drift net fishing, whaling, or habitat destruction, using photographic equipment. The group has been criticized, perhaps most strongly by Greenpeace, for going beyond the channels of negotiation, protest, and compromise when such actions prove ineffective.

This concern is echoed by the governments of Norway and Denmark, who have complained of illegal interference from Sea Shepherd personnel. In turn, the organization's founder points out that these nations conduct whaling operations in violation of international law. Sea Shepherd claims responsibility for scuttling a number of unmanned illegal whaling ships from these countries. "Especially needed," reads a pamphlet on Sea Shepherd volunteers, "are civil or former military personnel and individuals with submarine experience."

Sea Shepherd's firm stance on environmental enforcement has drawn heated criticism from Greenpeace, whose very name implies a more passive and respectable philosophy. No one can say Greenpeace's methods are ineffective: they are probably responsible for drawing more people into environmentalism than any other group on the planet, and their effect on local and international legislation has been crucial to numerous environmental concerns. On the other hand, even the most stringent policies and conventions governing wild cetaceans lack enforcement: there just isn't any funding available for whale cops. Illegal whaling is very much alive and well in the 90's, which is more than you can say for its victims. Sea Shepherd is the only organization, governmental or otherwise, which actively enforces the International Whaling Commission's ban on whaling in international waters.

Debating the validity of the two organizations' approaches to whale conservation is a tricky and subjective business, and in the end may be pointless. The two fill separate voids in environmental issues. Greenpeace works in the spotlight to affect policy, and Sea Shepherd works behind the scenes to enforce policy when nations and individuals choose to ignore it. As long as Sea Shepherd maintains its relatively low-profile track record (no convictions and no injuries caused by any of its members in its 18-year history) the group will most likely continue to hamper illegal whaling without damaging the reputation of environmentalists worldwide. Sea Shepherd's founder, Paul Watson, co-founded Greenpeace's save-the-whales efforts. He has some interesting opinions about the two groups' activities, which you can find by hyperlinking yourself to his page in Sea Shepherd's website at:

http://www2.seashepherd.org/orgs/sscs/essays/secondgp.html.

Involvement Opportunities

If you're interested in volunteering for Greenpeace, you should send in an application to join their "Activist Network", which you can get by writing to the Washington office. Although Greenpeace has been more active in lobbying and participating in international treaty negotiations in recent years, the group is still throwing itself in front of bulldozers and ships all around the world. Every day Greenpeace protesters are making a statement somewhere, whether it be rappelling down the side of ARCO headquarters to hang a giant anti-drilling banner, or chaining themselves to hoppers to prevent coal pollution in Canada.

In addition to protest activity, Greenpeace also needs lots of help with its campaigns. Be sure to send a check for a membership with the application (currently $30). Greenpeace also accepts Visa, American Express, and MasterCard donations, and if you'd like to continue your support throughout the year, they can arrange to make monthly deductions from your bank account. The Greenpeace website also offers membership info, press releases, reports, photos, short Quicktime movies, Action Alerts, "What You Can Do" information, and instructions on how to subscribe to various electronic mailing lists.

Sea Shepherd needs volunteers for office and campaign work as well as crew members on its small fleet of ships. Send $25 for a year's membership and request an application for volunteer work. Especially needed are experienced engineers, navigators, electricians, cooks, medical personnel and, as mentioned above, ex-military personnel and submarine crew members. Expect hard work with major job satisfaction for those who fit in. It's definitely a memorable experience.

International Wildlife Coalition

70 E. Falmouth Hwy.
East Falmouth, MA 02536
Phone: (508) 548-8328
Fax: (508) 548-1988
http://iwc.org
iwcadopt@unix.ccsnet.com

Overview

The International Wildlife Coalition, founded in 1983 by Dan Morast, is a nonprofit organization dedicated to "rescuing, nurturing, and protecting wildlife and wild habitat." IWC focuses on the exploitation of wildlife, the destruction of habitat, and the capture and display of animals by aquariums, zoos, and circuses. IWC is recognized as an official Non-Governmental Organization at the Convention on International Trade in Endangered Species, the International Whaling Commission, and the United Nations.

Conservation Efforts

The IWC has made quite a bit of progress since its inception in 1983. IWC helped establish and still coordinates the Cape Cod Stranding Network, and is responsible for training most of the network's volunteers to help respond to and rescue stranded marine mammals.

The Quixote is the IWC's 26-foot research vessel used for whale rescue, research, and protection projects. A "Whale Patrol" program helps monitor local populations of cetaceans on a year-round basis by documenting sightings and rescuing whales from net and line entanglements, boat collisions, ocean pollution, and other threats.

The IWC's activities span at least seven countries on six continents, and include animal rescue programs in Australia, Canada, Brazil, Thailand, and Sri Lanka, anti-poaching efforts in the stomping grounds of African elephants, and harp seal protection in Canada. The IWC participates in international debates which have a direct effect on global environmental policy. The group has been instrumental in a number of policy-making decisions, including the banning of international ivory trading.

Involvement Opportunities

While the IWC has no formal volunteer program in place, other than the Cape Cod Stranding Network team, its members have been heavily involved in letter-writing campaigns to protect wildlife. The IWC offers a variety of informative publications, teachers' information kits, and awareness campaigns. They also administer one of the largest whale adoption programs, which provides funds for the Quixote and other protective activities. The IWC offers memberships for donations of any amount ($20 suggested). The IWC accepts Visa and Mastercard donations. Donations can be mailed or, if you're hip, submitted on their website in the "membership" section.

Save the Whales

P.O. 2397
Venice, CA 90291
Phone: (800) 942-5365
Fax: (408) 394-5555
http://www.tmarts.com/savethewhales
stw@www.tmarts.com

Overview

"Save the Whales". How many of us were first drawn into the environmental cause by those three words back in the late 1970's? Concern over dramatically diminished whale populations throughout the oceans had reached a fever pitch in the early 70's, resulting in the passing of the Marine Mammal Protection Act in 1972, and the entry of the newly-founded Greenpeace into the anti-whaling cause in 1975. That same year, a 14-year-old Los Angeles resident by the name of Maris Sidenstecker II designed a T-shirt to be sold in support of anti-whaling organizations. Within two years, the phrase had been used by thousands of environmentalists, and "Save the Whales" became a nonprofit organization as well as a battle-cry.

Conservation Efforts

In the past two decades Save the Whales used letter-writing campaigns to support a number of environmental initiatives, and to help block ecologically harmful legislation. STW also campaigns to prevent the capture of wild dolphins and whales, and works to release animals in display facilities.

Involvement Opportunities

Save The Whales is not a direct action group, but relies on education programs, petitions, letter-writing campaigns, and public and media appearances to spread its messages and accomplish its goals. It supports no paid positions but relies heavily on volunteers to accomplish these tasks.

Other Programs

Memberships in Save the Whales are available, as are whale adoption programs. STW also reaches out to schoolchildren throughout the southwestern U.S. and Mexico through its Whales On Wheels educational programs and beach cleanups.

The Whale Museum

PO Box 945
62 First Street North
Friday Harbor, WA 98250
Phone: (800) 946-7227
Fax: (360) 378-5790
http://www.whale-museum.com

Overview & Conservation Efforts

The Whale Museum in Friday Harbor wears many hats.It is the only museum in the world devoted wholly to cetaceans, and offers educational programs and exhibitry about them. The museum also coordinates the San Juan County stranding network, and is responsible for recruiting volunteers, notifying proper agencies of stranded animals, and just generally knowing what the heck is going on. The museum acts as a clearing house of local cetacean information, interacting with organizations like the U.S. Fish and Wildlife, the National Marine Fisheries Service, and the Wolf Hollow animal refuge. The museum sponsors local cetacean research projects, most notably the Center For Whale Research's efforts to map out the populations of the Southern Resident Community of Pacific Northwest Orcas.

From time to time, Whale Museum personnel conduct research projects themselves, such as behavioral studies of local whales in response to boat traffic. So all in all, the Whale Museum is probably one of the most active cetacean-related facilities on the entire West Coast.

Involvement Opportunities

Although The Whale Museum has little to offer in the way of paid positions, they do make use of both volunteers and interns. Volunteering in an administrative capacity at the museum would provide valuable exposure to the researchers and projects underway in the San Juan County area. The museum also accepts applications from people wishing to get involved in stranding response, which in San Juan County is a collaborative effort between many local organizations. Aspiring biologists may apply for fellowships at the Whale Museum, or even approach them with a research proposal.

Other Programs

The museum offers a host of fun, educational programs to visitors of all ages. Sleepovers, single and multiple-day programs cater to ages five to adult. Museum educators lead tours of the exhibit hall, and lead Discovery Labs for scholastic and general interest groups. Group tours are offered year-round for a very reasonable fee. Of particular interest to adults is the Museum's Naturalist Training Program. The class presents participants with an overview of regional natural history and the biology, ecology, and conservation of local marine mammals. College credit is available through Western Washington University. Additional programs include outreach to local and regional schools, environmental workshops, and special events. Most activities take place in the spring and summer. Write, call, or visit the Museum's website for more information.

The Orca Adoption Program is a big source of income for the museum and its educational/conservation efforts. Like many adoption programs in zoological parks, participants choose a specific individual to sponsor, and are provided information about that animal's characteristics and history. Unlike the zoos' programs, however, The Whale Museum's patrons adopt wild animals. For some, being able to identify with an animal in the wild and following its progress and relationships over the course of years, or even decades, can provide a more direct and possibly more satisfying link to environmental conservation. A basic adoption costs $25, and includes an adoption certificate with an identification photo, a biography of the whale, and on annual issue of the museum's Orca Update, a newsletter providing new information on the orca populations.

For $20 more you get a membership to the museum as well, which includes unlimited admission, discounts on merchandise, and the museum's quarterly newsletter. They'll also toss in a $20 National Geographic video, "Killer Whales: Wolves of the Sea". Higher levels of adoption are also available. Call or write for a brochure, which includes a listing and brief description of all ninety-six whales.

World Conservation Union (IUCN)

Cetacean Specialist Group
Dr. Randall R Reeves, Chair
Okapi Wildlife Associates
27 Chandler Lane
Hudson, Quebec JOP 1HO, Canada
Phone: (514) 458-7383
Fax: (514) 458-7383
http://www.iucn.org
mail@hq.iucn.org

Overview

Only the IUCN's Cetacean Specialist Group understands the general situation of ALL the world's cetaceans. The CSG is a branch of the IUCN's Species Survival Commission (SSC). It consists of over sixty environmental biologists around the globe working towards the survival of cetaceans everywhere. One of more than a hundred specialist groups in the IUCN, its members join thousands of other Union members desperately trying to hold this planet's ecosystems together while the rest of us tear the place apart.

The IUCN itself is a huge organization with members in 125 countries. Founded in 1948, it seeks to "influence, encourage and assist societies throughout the world to conserve the integrity and diversity of nature and to ensure that any use of natural resources is equitable and ecologically sustainable." The Union is headquartered in Switzerland, with Regional and Country Offices in more than forty countries, and relies on the efforts of over 6000 expert volunteers in its project teams and action groups. The organization continues to decentralize as it grows, relying more and more on its expanding network of regional and country offices, located principally in developing countries.

Conservation Efforts

Although the Cetacean Specialist Group is a small part of the IUCN, it plays a very large role in global cetacean conservation. The CSG acts as an organizational catalyst in that it serves as a link between eager researchers and projects which need them. The CSG tries to catch populations and species of cetaceans which fall through the cracks in existing conservation efforts. Small cetaceans, in particular, have no real representation within the International Whaling Commission, which is currently the only globally recognized (or nearly so) authority on cetaceans. Consequently, small cetaceans are suffering from habitat encroachment even within the coastal waters of countries like the U.S. and Canada. In some nations this problem is compounded by culling operations, bycatch, and even overharvesting. (Harumphjapancoughcough.)

Since their 1988 release of the Action Plan for the Conservation of Cetaceans, the CSG has identified about 70 specific cetacean projects in need of assistance or initiation, most of them in less developed nations. About a third of the projects have been completed or are fully underway, about half still need either more funding or additional assistance, and the rest have been removed from the list or incorporated into other programs. CSG members often take a role in a project's development and implementation. One of the their main goals, however, is to find researchers and groups in the areas where conservation problems are actually occurring, and to make sure that these people get the technical guidance and support needed to address those problems on their own. In addition to identifying necessary areas of research and finding the talent and funding for them, the CSG also participates in the review of data from completed projects and sees that it reaches appropriate government and conservation groups.

Involvement Opportunities

The CSG provides no opportunities for involvement, nor does it distribute information to the general public. It does put out a bound edition of its 1994-1998 Action Plan, titled Dolphins, Porpoises, and Whales. This is a must-have for anyone interested in global cetacean conservation. Nowhere else will you find a treatment, brief as it is, of the status of every currently recognized species of cetacean. The booklet also covers each of the CSG's recommended projects, as well as current topics relevant to cetacean conservation.

Part Three

Dolphin and Whale Resources

- ►Professional Associations
- ►Government Departments
- ►The Marine Mammal Stranding Networks
- ►Dolphin-Assisted Therapy
- ►Interest Groups

7
Professional Organizations

This chapter actually covers two associations, a couple of societies, and an alliance. I considered throwing in a druidic cult just to break things up, by my editor is already about to gnaw her fingers off and I don't want to push her over the edge.

Associations give professionals a place to share ideas and information, and to address issues which affect their entire field. More importantly (to you) they bring a huge amount of information to their conferences about who's doing what and where. The conferences are great places to meet people, hear news, and collect lots of leaflets, buttons, and an occasional coffee mug. Memberships cost between $20 and $50 per year for students, a bargain for what you're getting.

Most of them print a journal of some kind, and all of them circulate newsletters. These publications teach you what the field is like, and what kinds of projects are underway. You'll occasionally find important career development opportunities, and sometimes job offerings get posted. Don't forget about the American Cetacean Society on page 218. They hold a biannual conference as well, and the next one is coming up in the winter of '98.

Acoustical Society of America

Animal Bioacoustics Technical Specialty Group
500 Sunnyside Boulevard
Woodbury, NY 11797-2999
Phone: (516) 576-2360
Fax: (516) 349-7669

Overview

What area of science could possibly feel a greater need to congregate in huge numbers and make lots of noise? The Acoustical Society of America is one of our country's oldest and most renowned scientific organizations. Its rich heritage began in 1928, when some enterprising scientist came up with the idea of a society devoted to acoustics.

He sounded his idea off of his colleagues who, after hearing him out, decided they liked the sound of it and echoed his sentiments. The idea reverberated throughout the industry, and in December of that year the ASA's first official meeting took place. With an initial membership of over four hundred members, the meeting ended on a good note, and set the tone for the organization's future growth.

The Society now has over seven thousand members in dozens of professions. Members of investigate topics ranging from physics to engineering and animal bioacoustics It is this last item which will be of particular interest to cetacean enthusiasts. Most researchers attempting to unravel the mysteries of cetacean sonar and vocalizations are members of the Society, and like the rest of its membership, they exchange ideas and information through annual meetings and a journal. The journal, which presents over 7000 pages of information each year, is available on a searchable CD-ROM for Macintosh and Windows as well as in print, making it a particularly valuable research tool.

Membership

Membership is open to anyone with an interest for $70 per year. Student memberships are also available, and are a bargain at $25. Both include a subscription to the Journal, and students are automatically sent the CD version.

Alliance For Marine Mammal Parks and Aquariums

103 Queen St.
Alexandria, VA 22314
Phone: (703) 549-0137
Fax: (703) 549-0488

Overview

There are thirty-two marine life parks around the globe which along with six professional members make up the Alliance For Marine Mammal Parks and Aquariums. The Alliance was formed in 1987, and since 1992 has been headquartered in the Washington DC area. It is essentially an organization formed by and for higher level marine park administrators as a vehicle for monitoring (and participating in) changes in marine mammal legislation and regulation. Whenever the Marine Mammal Protection Act comes up for review, or when APHIS, USDA, or U.S. Fish and Wildlife are reformulating their policies, or if the National Marine Fisheries service needs input on captive cetaceans, the Alliance will be there. By having an organized and reputable body of professionals and scientists to advise and cooperate with regulatory and legislative groups, this sort of arrangement ensures responsible and reasonable management for marine mammals on display.

They achieve this goal by working in two directions. They help ensure firm and adequate standards of care for the animals, and also advise on unnecessary and sometimes burdensome restrictions which are occasionally proposed. When the MMPA was being reauthorized in 1994, the Alliance successfully lobbied to give full authority for care and maintenance of the animals to APHIS. They also recommended common-sense changes in the transportation and export of marine mammals.

The Alliance tackles other issues related to marine mammal display through various committees. Education, communications, research, and stranding response committees address specific issues within the industry.

Membership

Membership in the Alliance is not open to the general public. The office in Alexandria might be able to provide hard-to-find answers to questions about display facilities in the United States.

American Zoo and Aquarium Association

7970-D Old Georgetown Rd.
Bethesda, MD 20814-2493
Phone: (301) 907-7777
Fax: (301) 907-2980

Office of Membership Services

Oglebay Park
Wheeling, WV 26003-1698
Phone: (304) 242-2160
Fax: (304) 242-2283

Overview

Founded in 1924 as the American Association of Zoological Parks and Aquariums, or AAZPA, the Association changed its name recently to the American Zoo and Aquarium Association, to produce the sleeker, more hip acronym "AZA" (though if you think about it, there's still an "A" missing in there someplace). The AZA is the zoological community's primary vehicle for centralized discussion and management of animal care and animal conservation issues. By providing the industry with a forum for discussing zoological issues, the AZA has been instrumental in the conservation "renaissance" which has transformed the industry in the past sixty years from simple attractions to defenders of wildlife. Their current membership includes more than one hundred and eighty zoos and aquariums in North America, from the Bronx Wildlife Conservation Park in New York to the Happy Hollow Zoo in San Jose, California.

The AZA's primary goal is the conservation of the world's wildlife, and is probably one of the few organizations in the world which is in a position to do something about it. The AZA's voice is heard by over a hundred million visitors a year through its member facilities, and in national and international forums for conservation. It is heard in Congress as well, where it assisted in the reauthorization of the Endangered Species Act and the establishment of the National Institute for the Environment. The AZA works closely with the US Fish and Wildlife Service, the National Marine Fisheries Service, and APHIS, providing support for and input on the regulation of animals in the US.

The AZA assembled an army of conservation groups and committees to accomplish its conservation goals, which include

Scientific Advisory Groups, Fauna Interest Groups , Taxon Advisory Groups, and Species Survival Plans. The Scientific Advisory Groups (SAGs) consist of zoo-based and university scientists focused on a particular topic area, like "Reintroduction", "Behavior and Husbandry", or "Small Population Management". SAG members serve as technical advisors to the AZA Board of Directors and other AZA groups, and also form liaisons with university scientists working on topics of interest to the zoo community. A Fauna Interest Group (FIG) is an AZA committee designed to facilitate conservation programs of AZA member institutions in specific geographical areas, such as Brazil, Madagascar or Zaire. Taxon Advisory Groups (TAGs) attempt to coordinate AZA efforts to ensure responsible management of an entire taxon. (i.e., "Mammals", or "Invertebrates".)

Finally, the AZA's well-known Species Survival Plan (SSP) is a cooperative breeding and conservation program. More accurately it is a collection of programs, involving the coordinated efforts of over a hundred zoos and aquariums throughout North America. SSPs are designed to maintain a genetically viable and demographically stable population of a species in captivity. The can be the last line of defense in animal species conservation, and give a species a chance of being resurrected after we've clubbed it to death. The SSP has been responsible for the reintroduction of several "extinct" species to the wild, including the California condor, the black-footed ferret, and Przewalski's horse (the world's last true wild horse.)

Let's take a look, to pad out the book a bit, at the animals currently covered under the SSP. The program seeks to preserve addaxes, babirusas, barasinghas, bonobos (aren't those little French chocolates?) cheetahs, chimpanzees, black-and-white colobuses, African wild dogs, drills, circular saws, African and Asian elephants, black-footed ferrets, gaurs, gibbons, lowland gorillas, Rodrigues' fruit bats, Rodrigues' fruit salads, pygmy hippopotami, Asian wild horses, and jaguars.

Also, sun and spectacled bears, tree kangaroos, (though personally I don't see how saving just tree of dem will help) black lemurs, mongoose lemurs, ring-tailed lemurs, ruffed lemurs, chocolate-fudge lemurs, clouded leopards, snow leopards, lions, Siberian tigers, and sloth bears (oh my.)

Then there are the mangabeys...er, gangmabeys....no, bangmanges...(ah, who cares?) Goeldi's monkeys, okapis, orangutans, Arabian oryxes, scimitar-horned oryxes, Asian small-clawed river otters, New York small-brained cab drivers, giant pandas, red pandas, Chacoan peccarys, black rhinoceri, greater one-horned Asian rhinoceri, greater two-chinned Asian sumo wrestlers, Sumatran rhinoceri, white rhinoceri, cotton-top tamarins, and golden lion tamarins.

Finally, the program also covers maned wolves, Mexican gray wolves, red wolves, Grevy's zebras, Hartmann's zebras, Thomas' zebras, (I got them from a market in Basrah, and they're not looking at all well) palm cockatoos, Andean condors, California hooded cranes, red-crowned cranes, wattled cranes, white-naped cranes, construction cranes, Micronesian kingfishers, Bali mynahs, St. Vincent parrots, thick-billed parrots, Congo peafowls, Humboldt penguins,

(gasp)

Mauritius pink pigeons, pink lawn flamingos, great hornbills, not-so-great hornbills, Guam rails, cinereous vultures, Chinese alligators, Dumeril's ground boas, Dumeril's ground beef, Virgin Islands boas, Cuban crocodiles, Aruba island rattlesnakes, Puerto Rican crested toads, radiated tortoises, irradiated tortoises, haplochromine cichlids, and last and quite possibly least, partula snails.

Membership

I might have been kidding about a few of those, but you get the idea - saving the Earth's animals is an enormous job, and the AZA is doing more than its fair share. Although by nature the AZA is an organization run for and by Zoological professionals, anyone may obtain an AZA membership at the "Associate" level. This entitles you to the AZA's periodic newsletter, Communiqué, and a discount on AZA publications and registration for the annual AZA conference. For information on how to become a member, contact the AZA membership office.

International Marine Animal Trainers' Association

1200 S. Lake Shore Dr.
Chicago, IL 60605
Fax: (312) 939-2216

Overview

The International Marine Animal Trainer's Association is like the AMA of the dolphin training world. A vehicle for the exchange of information and the discussion of professional issues, IMATA is both an indispensable tool for marine mammal professionals and an excellent source of information for animal care freaks hoping to start a career in marine mammal husbandry and training.

The two most obviously useful tools provided by IMATA are its quarterly magazine and its annual conferences. The magazine, Soundings, lists topical information about marine mammal display facilities, presents insightful solutions to training and health care issues, discusses current issues facing members of the profession, and even lists job and internship postings.

The annual conference, usually held in October or November, is attended by marine mammal professionals from all over the world. It's a terrific place to make personal contact with representatives from potential employers, and the papers, panels, and poster presentations provide a concentration of up-to-date information on marine mammal subjects found nowhere else.

Membership

An associate membership in IMATA costs $60 per year, which includes a subscription to Soundings and a discounted registration fee for the conference. Student memberships are available for $40, and carry the same benefits.

The Society for Marine Mammology

c/o Terrie M. Williams, Secretary
Department of Biology
EMS Bldg. A-316
University of California
Santa Cruz, CA 95060
Phone: (408) 459-5123
Fax: (408) 459-4882
http://pegasus.cc.ucf.edu/~smm
williams@biology.ucsc.edu

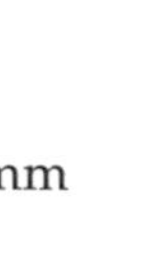

Overview

The Society For Marine Mammology is just as useful to aspiring cetacean scientists as the International Marine Animal Trainers' Association is to those wishing to be dolphin trainers. The SMM publishes a scientific journal, circulates a more informal newsletter, and holds professional conferences every two years.

The Society's journal, Marine Mammal Science, publishes more scientific findings concerning cetaceans than any other single publication. Not only is it a terrific place to keep abreast of recent scientific findings, but it is an excellent method of tracking down researchers with similar interests to your own. The Journal is stocked by many major public and university libraries, and a subscription is included in a Society membership.

The SMM also publishes the handbook, Strategies for Pursuing a Career in Marine Mammal Science, easily the most helpful publication on the subject to date. (Well, until this book came along, anyway.) The pamphlet revolutionized the almost nonexistent area of marine mammal career literature when it was first released in 1994. It offers a slightly different perspective on marine mammology careers than mine, and is definitely worth checking out. Copies may be ordered from the Society, but a free updated version is maintained on their website.

SMM conferences are treasure houses of current information about cetacean research, and are a great place to meet researchers and potential graduate advisors. The next conference, which may very well have already happened by the time you read this, is being held in conjunction with the Annual European Cetacean Society meeting, and is being billed as the "World Marine Mammal Science Conference". This huge gathering is being held on January 20th-24th in the Principality of Monaco, and should be quite an event. If you missed it, don't fret; the next conference is just one year later and is being held in Hawaii.

Membership

A student membership costs $50. Students have to prove their full-time status. Associate memberships will run you $75, but you still won't be able to vote. (Well, you can vote, but it won't count.) Memberships correspond to the calendar year, so if you join in April, they'll send you the journals from January-March. A lifetime membership in the Society is a real bargain at $1,200. Members receive the Society's membership directory, which is published every two years. (Hint, hint. MEMBERSHIP directory. Hmm. Could be useful.)

8
Government Departments

Somebody has to keep an eye on all of us (don't they?) and The Man can handle dolphins and whales just as well as our taxes. (Well, hopefully better than our taxes.) Two government departments split responsibility for cetaceans in the U.S. The Animal and Plant Health Inspection Service monitors display facilities while NMFS takes care of business in the wild. The Marine Mammal Commission keeps tabs on both types of animals, and passes on the information it gathers to APHIS & NMFS, the President, Congress, and anyone else who needs it. I listed the National Marine Sanctuaries Program since the waters they encompass harbor numerous cetaceans, and NMSP personnel often get involved in cetacean research and conservation. The International Whaling Commission isn't government per se, but certainly fits this category best.

Only NMFS and the NMSP offer volunteer opportunities. Both organizations are huge, and maintain many offices around the country. You'll have to track down the ones with volunteer programs yourself. No jobs exist with the IWC or the Marine Mammal Commission, and none of the employees at APHIS focus specifically on cetaceans. NMFS is your greatest government resource if your interest lies in cetacean research. It conducts cetacean research, commissions studies from other groups, and coordinates the stranding networks. NMFS's National Marine Mammal Laboratory in Seattle is a priceless resource for its library, which contains more information on dolphins and whales in one spot than anywhere else on Earth.

Animal and Plant Health Inspection Service

Animal Care
U.S. Department of Agriculture
Unit 84
4700 River Road
Riverdale, MD 20737
Phone: (301) 734-4980
Fax: (301) 734-4978
http://www.aphis.usda.gov

You might very well associate the United States Department of Agriculture with hog cholera, colostrum management, and outbreaks of the noxious Water Spinach (no kidding, look it up!) but the full scope of its responsibilities reaches much further. One of its departments, the Animal and Plant Health and Inspection Service has a great deal to do with cetaceans. For more than twenty-five years APHIS has enforced the Animal Welfare Act to protect certain animals from inhumane treatment and neglect. In particular, the Act requires that minimum standards of care and treatment be provided for certain animals bred for commercial sale, used in research, transported commercially, or (and here's the clincher,) exhibited to the public. That includes zoos, aquariums, or any other facility housing cetaceans.

Before APHIS will issue a license to display cetaceans, a potential applicant must be in compliance with all standards and regulations under the AWA. This includes very specific requirements concerning housing, handling, sanitation, nutrition, water, veterinary care, and protection from extreme weather and temperatures. Although Federal requirements establish acceptable standards, regulated businesses are encouraged to exceed the specified minimum standards, which they usually do.

Once a facility has been registered it is inspected by APHIS at least once a year, and more often if required. Random, unannounced checks by APHIS Veterinary Medical Officers and Animal Health Technicians ensure a high level of compliance, and generally keep staffs on their toes. APHIS, in conjunction with other regulatory agencies and the facilities themselves, does an excellent job of maintaining standards well above the global average, producing a national standard of animal care of which we can be quite proud. (Anyone interested in reading the fairly extensive minimum requirements set forth by the AWA should take a gander at Subpart

E of Part 3 of Subchapter A of the Act. The document is available from the main office, or at:

http://www.aphis.usda.gov/ac

You may also be interested to know that research facilities are required to establish an animal care and use committee to oversee the use of animals in experiments. The committee must be composed of at least three members, including one veterinarian and one person who is not affiliated with the facility in any way.

APHIS has just over 70 inspectors divided into three main regions (Eastern, Central, and Western,) and who are assigned specific facilities to inspect. To a certain degree it's the luck of the draw which determines whether or not they inspect cetacean facilities. As with nearly any Federal job, anyone wishing to apply for a job with APHIS must do so through the Office of Personnel Management. Check out the OPM website at :

http://www.opm.gov

International Whaling Commission

The Red House
Station Road, Histon
Cambridge CB4 4NP
United Kingdom

The International Whaling Commission was established in 1946 and was the result of an international convention on the subject of whaling held in Washington D.C. The organization's role has changed subtly since then, but its primary goal remains the same. It establishes a sustainable balance between whale conservation and commercial harvesting. Most of you should be aware that the "great whales" were driven to near extinction by the commercial whaling industry in the first half of this century, resulting in the near-extinction of most species of large whales. A great many of the nations responsible for this over-exploitation desire to repair the damage done, and are working very hard to make that desire a reality.

The IWC has just under 40 member nations, each of which appoints a commissioner to represent itself at the Commission's Annual and committee meetings. The organization is headquartered in Cambridge, England, where it maintains a Secretary, an Executive Officer, a Scientific Editor, a Computing Manager, and a small support staff. The annual meeting is held in May or June. In addition to the authority agreed upon my the member nations and established by the original 1946 Convention, the IWC is the recognized authority on whaling in the eyes of the United Nations, under Agenda 21 of UNCED (the United Nations Conference On Environment And Development Collection, 1992.)

While the IWC might have been viewed as more of a neutral body in its early years, recent times have seen a slight shift towards the conservation side of the whaling issue. In 1982, concerned with management difficulties and gaping holes in accurate whale population and sustainability data, the IWC voted to ban whaling worldwide until its policies could be responsibly reviewed and upgraded. The project was to take no longer than eight years.

Cataloging the world's whale populations, and planning an effective method of sustainable whaling has proven to be a tricky task. As of this writing, the global moratorium is still in effect. The IWC managed to establish figures for perhaps 40% of the target populations, so it's easy to see how painstaking population studies of cetaceans are.

In addition to gaps in critical data, the IWC faces great difficulties in policy development. The first step, the adoption of a new management procedure for any given population, was effectively solved at the 1994 annual meeting when the IWC embraced what is probably the most reasonable formula devised which would be accepted by the political crowd. The new policy's aim is to "obtain the highest possible continuing yield, with stable catch limits, to bring all stocks to 75% of their original level." It will ensure that depleted stocks are rehabilitated, and will prohibit any whaling on stocks below 54% of their initial level. IWC scientists have devised a reasonably accurate method of determining an "initial level" figure using only a stock's current size and accurate records of its catch history. Compared to the slaughter that was the whaling industry of the early 1900's, as well as the current status of baleen whales, this is a very reasonable approach.

No system is perfect, however, and for the above plan to work, the IWC will have to design an effective method of supervision and control to which all member nations will agree, a goal which has not yet been realized. The IWC is also in the process of addressing peripheral issues as well, including humane killing methods, aboriginal subsistence whaling, the creation of cetacean sanctuaries, whalewatches, and the IWC's jurisdiction over small cetacean-related issues.

When (or if) the ban on whaling is lifted, it will be interesting to see which countries will bother to resume or initiate commercial whaling. A few countries, most notably Norway and Japan, are already violating the ban, either in spirit through the misuse of "scientific" permits, or through an outright refusal to comply with regulations. In a vast majority of member nations, whales seem to be viewed as special in some way, and I doubt if very many of them would even resume whaling if given the chance to do so.

In any case, the IWC does not offer any educational, volunteer, or internship opportunities for the general public. It does, however, respond to public opinion in its decisions, so if you have something to say about whaling issues, send a request to your senator and representative to pass it along to the U.S. Commissioner.

Marine Mammal Commission

4340 East-West Hwy. Room 905
Bethesda, MD
Phone: (301) 504-0087
Fax: (301) 504-0099

"Whud" is exactly the sound you get by plopping the Annual Report to Congress of the Marine Mammal Commission onto a table. Though there are only three Commissioners, there are about two dozen people directly involved in the Commission's activities, and the content of their Annual Report, if somewhat federal in its tone, is extensive. In fact, a good solid perusal of the thing will tell you pretty much anything you want to know about marine mammal issues as they relate to our country and its government.

The Marine Mammal Commission is an independent agency of the Executive Branch, and its commissioners are, unsurprisingly, appointed by the President. The Commission was established in 1972 under Title II of the original Marine Mammal Protection Act. Its purpose is to develop, review, and make recommendations on the actions and policies of all Federal agencies with respect to marine mammal protection and conservation, and to carry out its own research program.

By law, the three commissioners are always individuals with an extensive background in marine ecology and resource management. The administrative office in Washington maintains a staff of about ten, and the Commissioners appoint a nine-member Committee of Scientific Advisors, who must be knowledgeable in marine ecology and marine mammal affairs. Together, this merry band is charged with overseeing the well-being of all marine mammals within the United States and its waters, and monitoring and advising all government agencies which regulate, protect, or encounter marine mammals. In addition, the Commission is responsible for keeping Congress (and presumably the Executive Branch as well) apprised of all marine mammal issues through its annual report.

The Commission's 1996 Annual Report starts out by highlighting species of special concern, such as manatees, Hawaiian monk seals, bowhead whales, etc. Then it addresses issues between marine mammals and the fishing industry, followed by conservation, Arctic animals, and strandings. It discusses pollution, oil and gas development, and federally funded research. Finally, the report looks at permits and authorizations to capture or interact with marine mammals in the wild, and ends with a discussion of several issues relating to the display industry. The report is supplemented by a number of appendices, tables, and figures.

It's a big job.

Although there are obviously very few, if any, opportunities for entry-level involvement with the Commission's office, it does regularly respond to requests for information; simply write to the above address. Unlike most of the organizations mentioned in this book, the Commission is not easily researched through libraries or the internet, and I would recommend consulting a recent copy of an Annual Report for further information.

National Marine Fisheries Service

Wild cetacean issues are handled by the National Marine Fisheries Service, or NMFS. ("NMFS", is pronounced just like the word "nymphs". If you don't know how to pronounce "nymphs", you should take a break from your job search and invest in a dictionary.) NMFS is a part of the National Oceanic and Atmospheric Administration. (NOAA is pronounced just like "Noah", the guy who wrote that dictionary you're going to buy.) NOAA, in turn, is a part of the Department of Commerce, and is an enormous organization with many subordinate groups and even its own fleet.

Under the Marine Mammal Protection Act of 1972 , the National Marine Fisheries Service's Marine Mammal Division is responsible for the protection and conservation of all whales, dolphins, porpoises, seals and sea lions. (Hereafter the term NMFS will refer to the Marine Mammal Division, unless otherwise noted.) NMFS scientists establish an "optimum level" for each species, and monitor their population levels. If a population falls below its optimum level, it is classified as "depleted," and a conservation plan is developed to guide research and management actions to restore the population to healthy levels. Endangered or threatened marine mammal populations also receive protection under the Endangered Species Act. (Eleven species of marine mammals in U.S. waters are either threatened or endangered right now, including most of the large whales.)

NMFS also acts as a buffer between marine mammals and the commercial fishing industry, and manages subsistence hunting of whales by Alaskan natives. NMFS is required under the MMPA to submit an Annual Report to Congress on the administration of the Act, and the Office of Protected Resources publishes a bi-monthly MMPA Bulletin detailing efforts to implement the MMPA.

Office of Protected Resources
1335 East-West Highway
Silver Spring, MD 20910
http://kingfish.ssp.nmfs.gov/tmcintyr/prot_res.html

Phone: (301) 713-2332
Fax: (301) 713-0376

The OPR is one of six "Headquarters" offices within NMFS, and is the one primarily responsible for accomplishing NMFS' goals in the area of marine animal conservation. Acting in tandem with Regional Offices and Science Centers, the OPR coordinates marine animal research and management efforts for NMFS.

Marine Mammal Division
1315 East-West Highway,
Silver Spring, MD 20910-3226
Phone: (301) 713-2322
Fax: (301) 713-0376

Within the OPR, the Marine Mammal Division handles all issues specific to marine mammals. The MMD works in conjunction with regional stranding networks and the Smithsonian Institute (see separate entry) to coordinate stranding efforts, and maintains the National Marine Mammal Tissue Bank. This bank preserves tissue in liquid nitrogen for future contaminant analysis, a valuable tool in establishing the health of animal populations.

Regional Offices and Science Centers

While the OPR and MMD central offices are left to handle "higher level" issues and decisions, the groundwork for NMFS' conservation and management plans is carried out through its many field offices. Handling the administrative and managerial tasks are the Regional Offices, and carrying out the necessary research are the Science Centers. The two work very closely with one another, and have in fact been combined in two of the five regions. The Regional Offices are good sources of information on management issues, but it is in the Science Centers and their subordinate laboratories that opportunities to get involved can be found.

Certainly most and probably all opportunities to get involved with NMFS research will be on a volunteer basis. Already contending with the general slump in funding for marine mammal research, NMFS has also felt the effects of government cutbacks and drawdowns, and has little money to spare. In addition, there seems to have been a bit of a hiring freeze in many of NMFS' departments in 1996, and it is impossible to predict how long, or if, it will continue. Nevertheless, the various Science Centers continue to use volunteers for lab work, shore and ship-based observations, and photo identification work, and once in a while a few of them do get hired on. Any chance to work with NMFS' marine mammal scientists is well worth the effort, simply because no other organization has as much experience in cetacean conservation.

Bear in mind also that NMFS may choose to task out certain research projects to local private or university researchers, probably researchers with a history of working with NMFS. If the NMFS personnel you contact aren't conducting cetacean research themselves, perhaps they can direct you to someone who is.

Alaska Regional Office
709 9th St.
Juneau, AK 99801-1807
Phone: (907) 586-7221
Fax: (907) 586-7249

In addition to its non-marine mammal tasks, the ARO also handles permits for local fishermen whose target catches bring them into contact with marine mammals, and publishes a guide to viewing marine mammals in Alaska.

Alaska Fisheries Science Center
7600 Sand Point Way, NE Bldg. 4
Seattle, WA 98115-0070
http://columbia.wrc.noaa.gov/afsc/home.html
Phone: (206) 526-4172
Fax: (206) 526-6615

Home to the National Marine Mammal Laboratory (which does not work with captive cetaceans) the AFSC once headed up nearly all cetacean research within NMFS. Although the individual Science Centers have assumed some of that responsibility in recent years, the AFSC still handles a fair amount of cetacean research. The Center is essentially responsible for fisheries research in the coastal waters off Alaska and the West Coast of the United States, including the northern Pacific Ocean and the eastern Bering Sea. It's a big task, encompassing some of the most important commercial fisheries in the world (like walleye pollock, Pacific salmon, and King and Tanner crabs). Useless fact freaks will be impressed that this constitutes a total biomass of more than 26 million metric tons and yields 65% of the volume and about 50% of the value of the total U.S. domestic catch in the nation's 200-mile Exclusive Economic Zone. The largest marine mammal populations in the nation are in the same area; which is probably the reason for locating the National Marine Mammal Laboratory there.

National MM Laboratory
7600 Sand Point Way, NE Bldg. 4
Seattle, WA 98115-0070
http://nmml01.afsc.noaa.gov/
Phone: (206) 526-4047
Fax: (206) 526-6615

The NMML may no longer be the national hub of marine mammal research it was when created in 1978, but it does have the single most valuable library on the planet for marine mammal research. Nowhere else is there such an extensive and complete collection of marine science journals, newspaper clippings, books, and stuff about marine mammals. There's this whole bank of filing cabinets in there that contains an alphabetical listing of pretty much everyone who's ever done a study of some kind on marine mammals, with copies of each study tucked inside. I just kind of stood there, drooling, and seriously considered moving to Seattle until this book was finished. It's open to the public, so if you ever wind up in Seattle, be sure to pay a visit.

The NMML is currently running five main programs. The first four don't really involve much cetacean research. The Alaska Ecosystem Program is primarily responsible for advising on the status of Steller sea lions, northern fur seals and harbor seals. The Antarctic Ecosystem Program conducts pinniped and seabird research which has been identified as "high priority" by the CCAMLR. (I'm not going to elaborate on that acronym - they're driving me nuts. Suffice it to say that the CCAMLR is an agreement between the Antarctic Treaty nations (which includes the US) to keep us all from screwing up Antarctica.) The California Current Ecosystem Program is responsible for assessing marine mammals in Washington and Oregon as well as the direct and ecological interactions with fisheries along the coasts of Washington, Oregon, and California. The Ecosystem Ecology and Assessment Program develops and provides scientific advice on ecosystem management with regard to management of renewable resources.

The one program which does involve a fair amount of cetacean work is the Arctic Ecosystem Program, which is being run by Dr. Doug DeMaster. This project is charged with determining the status of a number of cetacean and pinniped species including bowhead, beluga, gray, killer and humpback whales, and harbor porpoise. There is a chance that a very limited number of volunteers could become involved with this project, perhaps assisting with photo identification, field work, or observations. Direct queries about the Arctic program to Dr. DeMaster.

An additional Lab under the AFSC which is not currently involved in cetacean research is in Auke Bay, AK - (907) 586-7221.

NE Regional Office Phone: (508) 281-9278
One Blackburn Drive
Fax: (508) 281-9371
Gloucester, MA 01930

NE Fisheries Science Center & Woods Hole Lab
Phone: (508) 548-5123
Fax: (508) 548-5124
166 Water St.
Woods Hole, MA 12543-1097
http://www.wh.whoi.edu/nefsc.html

The Protected Species Branch of the NEFSC uses ship and plane-based observations, mark and recapture techniques, and electronic data recording to keep abreast of population size and dynamics for the marine mammals of the northeast coast, including humpback and right whales. Particular attention has been given to the harbor porpoise in an attempt to reduce the number of incidental catches in gill nets. Satellite tagging, genetic analysis, and acoustic alarms have all been or will be used by Branch staff members to that end. The Center has been able to take on a limited number of volunteer interns, research cruise volunteers, and student biological aides in support of these programs. Contact David Potter at extension 262 for more information.

Additional labs under the NEFSC which are not currently involved in cetacean research include:

Gloucester, MA Phone: (508) 993-9309
Narragansett, RI Phone: (401) 782-3200
Milford, CT Phone: (203) 783-4200
Sandy Hook/Highlands, NJ Phone: (908) 872-3000
Washington, DC Phone: (202) 357-2550

NW Fisheries Science Center, NW Regional Office & Montlake Lab
Phone: (206) 860-3200
2725 Montlake Blvd. E.
Seattle, WA 98112
http://research.nwfsc.noaa.gov/index.html

Since most cetacean issues and research for the West Coast are handled through the AFCS, little cetacean work is done either at the NWFSC or its subordinate laboratory in Portland, OR.

SE Regional Office Phone: (813) 570-5328
9721 Executive Center Drive North
St. Petersburg, FL 33702

SE Fisheries Science Center, Phone: (305) 361-4284
Miami Lab, & Research Mngmt. Div. Fax: (305) 361-1219
75 Virginia Beach Dr.
Miami, FL 33149
sefscweb@ccgate.ssp.nmfs.gov

The SEFSC is doing a great deal of cetacean population and distribution studies, and it should be possible for an enterprising volunteer to get in on the activity. The Center is also investigating regional mass strandings of recent years, attempting to learn more about their cause and at the same time to improve stranding response and sampling techniques. The Miami laboratory also coordinates stranding response in the area, as well as for the entire SE Marine Mammal Stranding Network.

Oxford Laboratory Phone: (410) 226-5901
904 S. Morris Street Extended
Oxford, MD 21654

The Oxford Lab is cooperatively run by NMFS and the Maryland Department of Natural Resources, and was only recently transferred from the NEFSC to the SEFSC. There are several biologists in the Oxford Lab who respond to cetacean strandings and conduct necropsies, including toxicology and histology examinations. They would like to implement some kind of volunteer program, so if you're interested, contact Joyce Evans at the above number.

Charleston Laboratory Phone: (803) 762-8500
217 Ft.Johnson Rd. Fax: (803) 762-8700
P.O. Box 12607
Charleston, SC 29412

The Charleston Lab has been conducting the analyses of stranded animals in the SEFSC's efforts to determine the cause of recent mass strandings. The Lab has conducted a number of cetacean research and health monitoring programs, and even offers internships on a limited basis. The Lab responds to marine mammal strandings, and utilizes volunteers in its laboratory work.

Galveston Laboratory
4700 Ave. U
Galveston, TX 77551
Phone: (409) 766-3500
Fax: (409) 766-3508

Although Galveston's focus is on sea turtles, the lab is also evaluating the impact of oil rig removal on marine mammals, among other species.

Mississippi Laboratories
3209 Fredric St.
Pascagoula, MS 39567
Phone: (601) 762-4591
Fax: (601) 769-9200

The Mississippi Labs are, among other things, responsible for constructing radio & satellite transmitter packages for tracking marine mammals.

Additional Laboratories under the SEFSC which are not currently running any cetacean-related programs are:

Beaufort, NC
St. Petersburg, FL
Panama City, FL
Bay St. Louis, MS

SW Regional Office
501 West Ocean Blvd, Suite 4200
Long Beach, CA 90802-4213
http://swfsc.ucsd.edu/swr.html
Phone: (562) 980-4000

The territory covered by the Southwest Region encompasses California, Hawaii, and the Pacific Trust Territories. The region also maintains administrative offices in Honolulu, Hawaii, American Samoa, and Santa Rosa and Eureka, California.

SW Fisheries Science Center &
La Jolla Laboratory
8604 La Jolla Shores Dr.
P.O. Box 271
La Jolla, CA 92038

Phone: (619) 546-7000
Fax: (619) 546-7003

The La Jolla Laboratory is located on the campus of the University of California, San Diego, and happens to be right next to the Scripps Institution of Oceanography. The whole complex is perched atop this amazing seaside cliff, and the staff members on the west side of the main building (which, like many NMFS office complexes, looks like a drive-in motel gone horribly wrong,) have one of the most breathtaking views in the world of Government Service.

The Marine Mammal Division, one of four divisions within the La Jolla Lab, is responsible for monitoring the status of cetacean populations in the eastern tropical Pacific and the coastal marine mammals of California. Using sea, aerial, and shore-based surveys the La Jolla staff monitors the population size, habitat use, and population dynamics of over a dozen species of marine mammals. The staff encounters such species as spotted, spinner and common dolphins, blue and fin whales, and various pinnipeds.

The lab occasionally steps outside the realm of population demography to pursue more physiological studies, such as the diving times of deep-diving whales in the Gulf of California. Other projects have utilized molecular biology, biogenetics, and forensic identification of unknown samples from stranded animals. Researchers and their volunteers also get to play with the military-donated thermal night sights the lab has laying around, so if you've always wanted to be a spy but don't like guns, this might be a good place to hang out. The Division's staff also develops and tests mathematical models to determine how factors like reproduction, growth, and geographic distribution of marine mammal populations might influence changes in population levels.

Anyone wishing to volunteer their assistance should contact one of the three research scientists most heavily involved in cetacean research: Drs. Jay Barlow, William Perrin, and Dave Perryman.

Honolulu Laboratory
2750 Dole St.
Honolulu, HI 96822-2396

Phone: (808) 943-1221
Fax: (808) 943-1290

The Honolulu lab is smack dab in the middle of the University of Hawaii's fairly large campus, which for you surfing gurus out there is about a ten minute drive from Waikiki beach. In addition to the main lab facility, NMFS researchers also have access to some facilities at the Kewalo Basin Marine Lab (see Dolphin Institute, page 61). NMFS/Honolulu is organized into five primary areas of research: marine mammals and endangered species, insular resources, fisheries oceanography research, and fishery management research. The marine mammal and endangered species group deals with the recovery of the Hawaiian monk seal, as well as local sea turtles.

Although almost no cetacean research is conducted at NMFS' Honolulu Laboratory, their field studies of endangered local animals provide an excellent opportunity for graduate students or recent graduates to get a firsthand look at conservation biology at work, and could very well open the door to other opportunities within the entire NMFS superstructure.

Accurately monitoring the monk seals' populations means sending NMFS personnel out to catch, measure, and tag as many of them as possible, as well as posting observers nearby to...well, to observe, really. Field teams of up to four or five people are sent out for anywhere from four to eight months at a time. Living conditions are pretty rugged, since the sites are scattered around some of Hawaii's more remote islands far to the northwest. Terrain tends to be pretty minimalistic on the deserted, dead coral islands, and often consists of little more than sand, water, and air. A team might be resupplied once during the trip, so pretty much everything they need is packed in from the beginning, and once the team is in place, they're committed for the long haul (barring emergencies.) Each day is spent observing, recording numbers, collecting debris, and tagging animals.

On the plus side, NMFS actually pays all the team members! Average salary for new team members is probably around $2000 per month before the infamous Hawaiian Tax Vampires suck out their cut, but since the field workers have no way of spending their paychecks while on assignment, things tend to balance out nicely.

There are only about nine full-time employees assigned to the monk seal project, so there is a healthy demand for temporary team members. NMFS tends to look for people who have already completed a bachelor's degree, and prefer someone who has an interest in marine mammology in some form, though they do take others. Most of them have only been out of college for a year or so, and some are simply taking a break from their graduate studies. Experience with PC compatibles (yuk!) is a plus, since all the data is entered while in the field. Volunteer experience in another field study, or even a zoo or aquarium, is also helpful.

National Marine Sanctuary Program

1305 East-West Highway, 11th Floor
Fax: (301) 713-0404
Silver Spring, Maryland 20910
Phone: (301) 713-3078
http://www.nos.noaa.gov/ocrm/nmsp/

Somewhere amongst your travels and research, you may have come across the term "National Marine Sanctuary". There are twelve of them here in the United States, and two more have been proposed. They're part of a government program, so get ready for some long names. The National Marine Sanctuary Program was established in 1972 through the Marine Protection, Research and Sanctuaries Act, and is administered by the Sanctuaries and Reserves Division of the National Oceanic and Atmospheric Administration (NOAA). The mission of the program is to "identify, designate and manage areas of the marine environment of special national significance due to their conservation, recreational, ecological, historical, research, educational" blah blah blah. Politicians must get paid by the word.

Basically, the sanctuary program identifies coastal or oceanic areas which are particularly valuable and sets up a local office to watch over it. These "satellite" offices also run educational programs and conduct or participate in research, often in conjunction with big marine labs, universities, or other government branches. Many of the local offices use volunteers or interns, and they could provide valuable exposure to conservation biology involving cetaceans. All can be contacted directly by snail mail, phone, fax, or Email. Here's a listing of the sanctuaries with brief descriptions of their programs. (Information furnished by NOAA.)

Channel Islands

National Marine Sanctuary
113 Harbor Way
Santa Barbara, CA 93109
Phone: (805) 966-7107
Fax: (805) 568-1582
Email: channel_islands@ocean.nos.noaa.gov

The Channel Islands National Marine Sanctuary encompasses the waters surrounding San Miquel, Santa Rosa, Santa Cruz, Anacapa, and Santa Barbara Islands. A fertile combination of warm and cool currents results in a great variety of plants and animals, including large near shore forests of giant kelp, flourishing populations of fish and invertebrates, and abundant and diverse populations of cetaceans, pinnipeds, and marine birds. Located about 25 miles off the coast of Santa Barbara, the sanctuary covers an area of 1,658 square miles, and was designated in September of 1980.

Educational activities run by the sanctuary office in Santa Barbara include internship and volunteer programs, classes and curriculum for teachers and students, interpretive exhibits, lectures, and events. Publications include newsletters, brochures, posters, an educational resource directory, and a research bibliography with over 4,000 scientific references. Sanctuary staff and volunteers frequently get involved in oceanographic, seabird, marine mammal, kelp forest, and intertidal research by local, state, and federal agencies. Sanctuary staff are currently investigating the unusually high number of blue whales which are congregating off the Northwest section of the sanctuary's Channel Islands. The research program is called WHAPS96 (which stands for Whale Habitat and Prey Study) and is a cooperative effort between the sanctuary, Scripps Institution of Oceanography, and the National Marine Fisheries Service's Southwest Fisheries Science Center.

Cordell Bank and Gulf of the Farallones
National Marine Sanctuaries Phone: (415) 561-6622
Fort Mason Building #201 Fax: (415) 561-6616
San Francisco, CA 94123
http://www.nos.noaa.gov/nmsp/gfnms
Email: cbnms@ocean.nos.noaa.gov

Cordell Bank is an offshore seamount where the combination of oceanic conditions and undersea topography creates a highly productive marine environment. The Bank rises to within 115 feet of the sea surface with water depths of 6,000 feet only a few miles away. The prevailing California Current flows southward along the coast and the upwelling of nutrient-rich, deep ocean waters stimulates the growth of organisms at all levels of the marine food web. It is a destination feeding ground for many mammals and seabirds, including Humpback whales, blue whales, and Dall's porpoises. The sanctuary is 60 miles northwest of San Francisco, covers 526 square miles, and was designated in May of 1989.

The Gulf of the Farallones sanctuary includes nurseries and spawning grounds for commercially valuable species, at least 26 species of marine mammals (including gray whales,) and 15 species of breeding seabirds. One quarter of California's harbor seals breed within the sanctuary. The Farallon Islands are home to the largest concentration of breeding seabirds in the continental United States. The sanctuary boundaries include the coastline up to mean high tide, protecting a number of accessible lagoons, estuaries, bays, and beaches for the public. The sanctuary spreads along the coast of California north and west of San Francisco, covers 1,255 square miles, and was designated in January of 1981.

The Ft. Mason office offers a number of programs for both sanctuaries. For Cordell bank, they run an educational unit on deep ocean ecology and pelagic ecosystems, and give public presentations on the sanctuary and related topics. They also offer teacher workshops and training. Research activities include pelagic surveys for seabirds and marine mammals, sampling for state biotoxin monitoring program, monitoring abundance and distribution of krill within sanctuary waters, and research on the feeding ecology of minke whales.

For the Gulf of the Farallones sanctuary, available programs include community outreach marine programs, seminars, presentations, and publications. The staff also runs Sea Camp, a week long summer camp for low income children. Current research concerns rocky intertidal habitat monitoring, oil spill monitoring, marsh restoration in Tomales Bay, and habitat restoration for common murre. Beach Watch is a volunteer program that trains the public to conduct biological monitoring within the sanctuary.

Fagatele Bay
National Marine Sanctuary
P.O. Box 4318
Pago Pago, Americam Samoa 96799
Email: fagatele@ocean.nos.noaa.gov
http://www.nos.noaa.gov/nmsp/FBNMS/

Phone: (684) 633-7354
Fax: (684) 633-7355

Fagatele Bay comprises a fringing coral reef ecosystem

nestled within an eroded volcanic crater on the island of Tutuila. Nearly 200 species of coral are recovering from a devastating crown-of-thorns starfish attack in the late 1970s, which destroyed over 90% of the corals. Since then, new growth has been compromised by two hurricanes, several tropical storms, and coral bleaching. This cycle of growth and destruction is typical of tropical marine ecosystems. The sanctuary is on the southwest shore of Tutuila Island, American Samoa, 14 degrees south of the equator. It is the smallest of the sanctuaries at a modest quarter of a square mile, and has been designated since April of 1986. The office runs a summer camp for students, school and village outreach programs, and a fourth grade program called ReefWeeks. They also mention on their web page that they produce, among other publications, a turtle coloring book, which is pretty darn cool. Research activities include coral reef habitat surveys, a temperature and water quality monitoring program, and a long term resource recovery survey.

Florida Keys
National Marine Sanctuary
P.O. Box 500368
Marathon, FL 33050
Email: fknms@ocean.nos.noaa.gov
Phone: (305) 743-2437
Fax: (305) 743-2357

The Florida Keys marine ecosystem supports one of the most diverse assemblages of underwater plants and animals in North America. Although the Keys are best known for coal reefs, there are many other significant interconnecting and interdependent habitats. These include fringing mangroves, seagrass meadows, hard bottom regions, path reefs, and bank reefs. Programs include ecosystem health monitoring, sanctuary-wide, multidisciplinary cruises to assess the current status of coral habitat, and onsite assistance for regional research projects. The sanctuary surrounds the entire Florida Keys archipelago, covers 3,674 square miles, and was designated in November of 1990. The Keys NMS runs educational training workshops and school programs, onsite interpretive tours, and a classroom and field study program for eighth graders.

Flower Garden Banks
National Marine Sanctuary
216 W. 26th St. Suite 104
Bryan, TX 77803
Email: flower_gardens@ocean.nos.noaa.gov
Phone: (409) 779-2705
Fax: (409) 779-2334

One hundred miles off the coasts of Texas and Louisiana, a pair of underwater gardens emerge from the depths of the gulf of Mexico like oases in the desert. The Flower Garden Banks are surface expressions of salt domes beneath the sea floor. This premiere diving destination harbors the northernmost coral reefs in the United States and serves as a regional reservoir of shallow water Caribbean reef fishes and invertebrates. Roughly 110 miles south of the Texas-Louisiana border, covers 56 square miles, and was declared in January of 1992. Flower garden NMS offers classroom and shipboard presentations as well as workshops for teachers and researchers. They also publish a biennial news bulletin and informative brochures, posters, and videos. Research programs focus on coral diseases, mass spawning, paleo-climatology, fish censuses, and elasmobranch surveys. (Personally, I'm considering commuting once a week from Chicago just to put "Elasmobranch Surveyor" on my resume.)

Gray's Reef
National Marine Sanctuary
10 Ocean Science Circle
Savannah, GA 31411
Email: grnms@ocean.nos.noaa.gov
Phone: (912) 598-2345
Fax: (912) 598-2367

Just off the coast of Georgia, in waters 20 meters deep, lies one of the largest near-shore sandstone reefs in the southeastern United States. The area earned sanctuary designation in 1981, and was recognized as an international Biosphere Reserve by UNESCO in 1986. Gray's Reef consists of sandstone outcroppings and ledges up to three meters in height, with sandy, flat-bottomed troughs between. Because of the diversity of marine life, Gray's Reef is one of the most popular sport fishing and diving destinations along the Georgia coast. A key species in the sanctuary is the Northern right whale. It lies 17 miles east of Sapelo Islands, Georgia, and covers 23 square miles. The sanctuary staff offers community outreach marine programs, seminars, and presentations as well as educational publications and exhibits. They are conducting a long term study of environmental conditions and processes within the sanctuary, as well as onsite projects by regional researchers, reef fish tagging and assessment, and archaeological surveys.

Hawaiian Islands Humpback Whale
National Marine Sanctuary
726 South Kihei Road
Kihei, HI 96753
Phone: (808) 879-2818
Fax: (808) 874-3815
Email: hihwnms@ocean.nos.noaa.gov
http://www.nos.noaa.gov/nmsp/hinms/

The shallow, warm waters surrounding the main Hawaiian Islands constitute one of the world's most important humpback whale habitats. Scientists estimate that two-thirds of the entire North Pacific humpback whale population migrate to Hawaiian waters each winter to engage in breeding, calving, and nursing activities. The continued protection of humpback whales and their habitat is crucial to the long term recovery of this endangered species. The sanctuary extends out to a depth of 100 fathoms surrounding the four island area of Maui, Penguin Bank, and off Kilauea Point, Kauai in Hawaii. It covers 1,300 square miles and was designated in November of 1992. The sanctuary offers classroom and shipboard presentations, a sanctuary newsletter, brochures, pamphlets, and runs a poster contest. They also conduct research and management workshops, study whale population and behavior, offer support for graduate student whale research, run a volunteer water quality monitoring project on Maui, and a Marine Option Program internship.

Monitor National Marine Sanctuary
c/o The Mariners' Museum
100 Museum Drive
Newport News, VA 23606
Phone: (757) 599-3122
Fax: (757) 591-7353
Email: mnms@ocean.nos.noaa.gov
http://www.cnu.edu/~monitor

Although it has little to do with cetaceans, it wouldn't be right to exclude a mention of the very first national marine sanctuary. The sanctuary surrounds the site of the wreck of the USS Monitor, a Civil War vessel that lies off the coast of North Carolina. The Monitor was the prototype for a class of U.S. Civil War ironclad, turreted warships that significantly altered both naval technology and marine architecture in the nineteenth century. Designed by the Swedish engineer John Ericsson, the vessel contained many of the emerging innovations that revolutionized warfare at sea. If you haven't read about it in school, you should check out the somewhat hilarious story about the titanic battle between the Monitor and the Merrimac (the South's answer to the ironclad Monitor.) Basically the two kept bouncing cannonballs off one another for days with no victory in sight for either of the invincible tubs. The sanctuary was designated on January 30, 1975, and covers a circle one mile in diameter, centered on the wreck, 16 miles south-southeast of Cape Hatteras, North Carolina.

Monterey Bay Marine
National Marine Sanctuary
299 Foam Street, Suite D
Monterey CA 93940
Email: aking@ocean.nos.noaa.gov
http://bonita.mbnms.nos.noaa.gov/
Phone: (408) 647-4201
Fax: (408) 647-4250

Monterey Bay, the nation's largest marine sanctuary, spans over 5,300 square miles of coastal waters off central California. Within its boundaries are a rich array of habitats, from rugged rocky shores and lush kelp forests to one of the deepest underwater canyons on the west coast. These habitats abound with life, from tiny plants to huge blue whales. With its great diversity of habitats and life, the sanctuary is a national focus for marine research and education programs. Sanctuary publications include natural history books, research and education directories, newsletters, brochures, posters, and videos. The staff offers educational programs, lectures and events for the general public as well as teachers and students. The sanctuary is developing research priorities with 20 local marine science institutions and encourages research through numerous avenues, including funding specific projects and disseminating scientific information through research publications and symposia.

Olympic Coast
National Marine Sanctuary
138 West First Street
Port Angeles, WA 98362
Email: ocnms@ocean.nos.noaa.gov

Phone: (360) 457-6622
Fax: (360) 457-8496

The Olympic Coast sanctuary spans 3,310 square miles of marine waters off the rugged Olympic Peninsula coastline. The sanctuary averages approximately 38 miles seaward, covering much of the continental shelf and protecting habitat for one of the most diverse marine mammal faunas in North America and a critical link in the Pacific flyway. The sanctuary boasts a rich mix of cultures, preserved in contemporary lives of members of Quinault, Hoh, Quileute, and Maka tribes. The sanctuary extends from Cape Flattery to the mouth of the Copalis River, on Washington's outer coast, covers 3,310 square miles, and was designated in July of 1994. Key species in the sanctuary include assorted species of dolphins and gray whales. Cultural programs are operated with the Makah, Quileute, Hoh, and Quinault Nations. Ecotourism programs are coordinated with state agencies and regional tourism organizations. The sanctuary staff conduct marine archaeological and biological studies, and other marine research coordinated with other federal and state agencies and universities. They also publish a quarterly newsletter.

Stellwagen Bank
National Marine Sanctuary
14 Union Street
Plymouth, MA 02360
Email: sbnms@ocean.nos.noaa.gov
http://vineyard.er.usgs.gov/

Phone: (508) 747-1691
Fax: (508) 747-1949

Formed by the retreat of glaciers from the last Ice Age, Stellwagen Bank consists primarily of coarse sand and gravel. Its position at the mouth of Massachusetts Bay forces an upwelling of nutrient-rich water from the Gulf of Maine over the bank--leading to high productivity and a multi-layered food web with species ranging from single-celled phytoplankton to the great whales. The sanctuary sits 25 miles east of Boston, 3 miles southeast of Cape Ann, and 3 miles north of Provincetown. It covers 842 square miles, and was designated in November of 1992. Northern right whales, humpback whales, and white-sided dolphins reside within the sanctuary's waters. The staff runs professional development workshops and an annual week-long "MimiFest" for grades 4-8, and have produced a sanctuary video and CD-ROM. The sanctuary's Aquanaut Program allows students to conduct research on benthic communities and acoustics through the National Undersea Research Center at the University of Connecticut.

9
The Marine Mammal Stranding Networks

The stranding networks. What a nightmare it's been to write about these things. No two organizations in them are alike, and even the networks themselves differ wildly in structure and organization. Nobody has a complete list of participating groups, except maybe me after tearing my hair out and running up a $500 phone bill. But hey, that's my problem, not yours. What you need to know is what these things are, and how you can get involved.

What are marine mammal stranding networks?

Each year, thousands of marine mammals wash up or strand on seashores all around the world, sometimes in large groups. Nobody knows why. Some believe whales are sensitive to the Earth's magnetic fields, and become confused when the fields fluctuate. Others suppose the animals follow leaders, who might lose their sense of direction due to illness or injury. Whatever the cause, somebody has to attend to these animals, whether alive or dead. Enter the stranding networks, stage right.

Stranding networks are loosely knit groups of people trained to respond at a moment's notice when dolphins or whales are found on a beach. These dolphin minutemen volunteer their time and efforts to help live animals, and learn as much as possible from dead ones. Each network covers an enormous expanse of coastline. This is a no-brainer, but in case anyone out there is a few bricks short of a wall, stranding response takes place along our seashores. This limits the networks to 22 states.

This book covers the four continental stranding networks:

Northeast Marine Mammal Stranding Network

Connecticut	Maryland	New York
Delaware	Massachusetts	Rhode Island
Maine	New Jersey	Virginia

Southeast Marine Mammal Stranding Network

Alabama	Louisiana	South Carolina
Florida	Mississippi	Texas
Georgia	North Carolina	

Northwest Marine Mammal Stranding Network
Oregon
Washington

California Marine Mammal Stranding Network
California (duh)

Complete information on the Alaska and Hawaii networks wasn't available at press time, so I'll have to do the phone number hand-off thing you see in other career guides. Here's who to call for more information:

Alaska Marine Mammal Stranding Network
Gene Nitta, Network Coordinator (808) 973-2987

California Marine Mammal Stranding Network
Kaja Brix, Network Coordinator (907) 586-7235

A NMFS office coordinates each network, though no two networks work the same way. The Southeast network gets the most animals, uses the most volunteers, and is very structured. By contrast, the Northwest network sees almost no cetacean strandings, uses very few volunteers, and is very loosely organized. Both the Northeast and California networks fall somewhere in between in all three aspects.

Organizations wishing to join a network apply for a Letter of Authorization from NMFS. A very few experienced professionals obtain LOAs for themselves, but most letters go to established organizations with a history of marine mammal work. Some aquariums and marine parks hold letters. So do some of the research and conservation groups listed in this book. (This chapter tells you which.) In all, more than sixty separate organizations participate in stranding response in the U.S.

What do network volunteers do?

What you get to do as a network volunteer depends heavily on which organization you're with. Some groups are more involved than others, and your geographic location has great bearing on how many strandings occur near you.

There are far more dead strandings in the U.S. than live ones. Volunteers responding to dead animals often take measurements and tissue samples from nasty, smelly carcasses. Tissue analysis and necropsies bring scientists closer to explaining why strandings occur. Skulls yield much valuable information in the hands of laboratory or museum personnel, so some cetacean corpses are even beheaded before disposal. Many volunteers involved in measuring and sampling carcasses find opportunities to assist in the analysis of the data and samples they take. Veterinary and pathology students may find this particularly valuable experience.

(Photo by Laurel Canty Ehrlich and provided by DRC)

If you'd be this happy stripping the flesh from a whale skull, you might have a future in stranding response.

Not all stranding work is so gruesome. Though a low percentage of strandings involve live animals, a fair number of dolphins and whales require care and attention each year. Some receive medical attention onsite, and get shoved right back out to sea. Many live for only a few hours. In such cases, volunteers try to keep the animals' blowholes above water, their skin wet, and crowds away. These ordeals involve long hours in the water, often in cold winds or blistering heat.

If a dolphin or a whale is too sick or injured to swim off on its own, yet survives long enough for a team to be assembled, it may be taken to a rehabilitation facility. About two dozen facilities in the U.S. can accommodate cetaceans to some degree, but many suffer from limited space and poor funding. Caring for a sick or injured cetacean can run into thousands of dollars in expenses each day! To make matters worse, a highly contagious and deadly disease called morbillivirus reared its ugly head in the early 90's, forcing many aquariums and marine parks to stop taking stranded animals.

When animals are rehabilitated at a marine park or aquarium, volunteers usually (though not always) help paid staff members care for them. These might be volunteers who normally give some of their time each week in a department, or they might be brought on especially for the rehabilitation effort. Some animals are rehabilitated in specially constructed ocean pens, or in temporary facilities like a marine laboratory. These kinds of efforts usually involve all-volunteer groups like Wildlife Rescue of the Florida Keys, or the Texas Marine Mammal Stranding Network.

In any case, rehabilitating an animal involves long hours and lots of hard work. Fish needs to be prepared, kitchens and pools need to be cleaned, and observations run throughout the night. Volunteers might even be involved, however rarely, in giving physical exams and tube-feeding. The whole process can drag on for days, weeks, or even months. Eventually the animal dies, gets well enough to return to the sea, or is deemed unfit for release and transferred to a permanent home. All three possibilities tend to hit the animal care team pretty hard. Rehabilitating any animals carries its own emotional baggage.

Are stranded animals the only ones needing rehabilitation?

Particularly in the Southeast and California Stranding Networks, a fair number of rescue and rehabilitation workers would like to see display animals released back into the wild. The debate over whether such releases are desirable or feasible lays outside the scope of this book (for now). However, a recent series of events in Florida shows us any such attempts require careful planning and collaboration to have any chance of success.

Until it lost its license to house cetaceans, Sugarloaf Marine Sanctuary in the Florida Keys cared for a number of bottlenose dolphins. The sanctuary was intended to be a rehabilitation center where dolphins from display facilities could be prepared for release. Luther and Buck, a pair of ex-Navy dolphins, had been sent to the facility against the Navy's recommendations, and at the behest of some congressmen who had been swayed by the Humane Society, which in turn had old ties with the facility's Director, Ric O'Barry.

After a time, the staff at Sugarloaf determined the two dolphins were fit for release, and requested a permit from the National Marine Fisheries Service. NMFS disagreed with the decision, and refused to issue the permits. On May 23rd, 1996, O'Barry released Luther and Buck into the waters of the Florida Keys anyway, under the assumption they were ready to face the challenges of life on their own.

Within days the dolphins were seen, seventy miles apart from one another, to be approaching humans in a manner suggesting they were begging for food. Luther had been approaching powerboats and jetskis, and sustained a propeller wound. Neither was seen in the company of other dolphins, and it quickly became intuitively obvious they would not survive for long on their own. Once a pair of rescue teams formed by NMFS, the Dolphin Research Center, and other members of the Marine Mammal Stranding Network managed to attract the dolphins' attention, they were given physical exams and found to be underweight, dehydrated, and both bore lacerations on their backs. Luther's skin, normally firm like a rubber boot, was unnaturally soft from malnutrition.

O'Berry maintained the animals' poor health was a "manufactured emergency" by the National Marine Fisheries Service, that Luther's injury was a rake mark, which is a harmless shallow scratch from another dolphin's teeth, and that both dolphins would have been fine if left alone.

It is far more likely that the National Marine Fisheries Service was simply doing its job, which was to investigate an illegal reintroduction of cetaceans into the wild. Only someone who had never seen dolphin rake marks could have looked at the wound on Luther's back and called it a rake mark, and the truth about leaving the dolphins out there on their own is simply that had it been done, they would now be dead.

The Sugarloaf incidents made it clear that releasing dolphins into the wild without adequate preparation is a very bad idea. There is evidence that reintroduction to the wild is feasible for many animals, but it should be done with the input and collaboration of experienced marine mammal care specialists from both the government and the private sector.

Where do I sign up?

Experience in the stranding networks is valuable for anybody interested in cetaceans. If you're looking for a career in animal care, research, or conservation, the experience will bolster your résumé, and give you contacts in the field. It's hard and sometimes unpleasant work which brings a great sense of satisfaction and accomplishment to those who stick it out.

Pick out three or four of the closest groups listed in this chapter and call them up. The descriptions on the following pages tell you what the group does, and doesn't do. Ask them if they're looking for volunteers at the moment. Find out what kind of commitment they ask for, and what kind of training they offer. Information in this chapter changes very rapidly. You can check the Omega website for updates, but calling groups directly is your best bet.

Stranding Network Descriptions

The Regional Networks are broken down by state, moving from north to south. Within each state the facilities are listed alphabetically. Some of the groups are also listed elsewhere in the book. In such cases I've made a quick note of how they are involved and whether they use volunteers, and then given the page number of the group's main description. Most of the other groups do stranding response as a sort of sideline, and bear fairly brief descriptions. A few, like The Marine Mammal Center in California, exist solely to care for stranded animals, which is why some descriptions seem unusually detailed.

Beluga whales make great filler for blank pages.

The Northeast Marine Mammal Stranding Network

Regional Coordinator:
Pat Gerrior
NMFS Woods Hole Lab
166 Water St.
Woods Hole, MA 12543-1097
Phone: (508) 495-2264
Fax: (508) 495-2258

Maine

Maine receives very few cetacean strandings each year. All of them are handled by members of Allied Whale at the College of the Atlantic (See below). Put these two facts together and you arrive at the conclusion that few opportunities for stranding response exist in this state. Try moving somewhere else.

Allied Whale
College of the Atlantic
105 Eden Street
Bar Harbor, ME 04609
Phone: (207) 288-5644
Fax: (207) 288-4126
alliedwhale@ecology.coa.edu

Allied Whale at the College of the Atlantic is the only authorized marine mammal stranding response organization in Maine. While AW has no formal volunteer program, they're still your only chance if you live in Maine. See page 151 for more information about Allied Whale.

Massachusetts

Far more dolphin and whale strandings occur in massachusetts than in Maine, especially along the Cape Cod area. Pilot whales in particular seem to have quite a bit of trouble avoiding that curved, squiggly thing at the end. In addition to five stranding response teams, this state offers one of the very few facilities in the Northeast capable of long-term rehabilitation, the New England Aquarium. Massachusetts also possesses the NMFS Woods Hole Lab, which coordinates regional stranding efforts. All of this makes the state a bit of a hotbed of activity.

Cetacean Research Unit
P.O. Box 159
Gloucester, MA 09130
Phone: (978) 281-6351
Fax: (978) 281-5666
http://www.friend.ly.net/user-homepages/b/birdman/index.htm
info@cetacean.org
Dave Morin - Intern Coordinator

While the CRU does respond to cetacean strandings, residents of the northernmost part of Massachusetts should contact the New England Aquarium (page 193). See page 165 for more information about CRU.

Center For Coastal Studies
P.O. Box 1036
Provincetown, MA 02657
Phone: (508) 487-3622
Fax: (508) 487-4495
http://www.provincetown.com/coastalstudies/index.html
ccswhale@wn.net

The CCS is very active in cetacean strandings and in disentangling whales caught in nets. Although the Center does enlist volunteer aid for such operations, initial training and contact for the Cape Cod Stranding Network is done through the International Wildlife Coalition. See page 156 for more information about the Center for Coastal Studies, and page 235 for the IWC.

International Wildlife Coalition
70 E. Falmouth Hwy.
East Falmouth, MA 02536
Phone: (508) 548-8328
Fax: (508) 548-1988
http://iwc.org
iwcadopt@unix.ccsnet.com

The IWC is a founding member of the Cape Cod Stranding Network, and handles volunteer recruitment and training for the entire region. See page 235 for more information about the IWC.

New England Aquarium
Central Wharf
Boston, MA 02110
Phone: (617) 973-5200
Fax: (617) 723-6207

The New England Aquarium is very active in both live and dead stranding response, and is capable of rehabilitating small cetaceans. Aquarium volunteers occasionally assist with these efforts. See page 193 for more information about NEAq.

Connecticut and Rhode Island

Strandings are less common in these states than in Massachusetts. The Mystic Marinelife Aquarium handles rescue and rehabilitation attempts for both states.

Mystic Marinelife Aquarium
55 Coogan Blvd.
Mystic, CT 06355-1997
Phone: (860) 572-5955
Fax: (860) 572-5969

The MMA handles all cetacean strandings in Connecticut and Rhode Island, and would be the best facility to approach about getting involved in those states. See page 108 for more information about MMA.

New York

99% of New York's strandings occur on Long Island, and are handled by the Riverhead Foundation.

New York State Marine Mammal and Sea Turtle Stranding Program
Riverhead Foundation for Marine Research & Preservation
431 E. Main St.
Riverhead, NY 11901
Phone: (516) 369-9840
Fax: (516) 369-9826
http://www.riverheadfoundation.org

My first impression of the rabid patrons of Manhattan's Penn Station one Friday afternoon was of a massive herd of lemmings headed mindlessly forward. Each weekend thousands of New York's more sensible residents evacuate their families to the comparatively rural sanctuary of Long Island. Living up to its name, the vast isle is a haven for beachcombers, sailors, bicyclists, and among other things, New York State's only marine mammal rescue and rehabilitation center.

Overview

The New York State Marine Mammal and Sea Turtle Stranding Program came into being in the 70's as the Okeanos Ocean Research Foundation, later incorporated in 1980 as a nonprofit research and educational center, and is now administered by the Riverhead Foundation for Marine Research & Preservation. Since then the group has built the Marine Mammal and Sea Turtle Stranding Program of N.Y. State from the ground up, providing assistance and care to over 150 animals a year. The Program has nursed turtles, pinnipeds, and cetaceans back to health, and is the only organization in the world to have successfully rehabbed and released a sperm whale!

The Foundation moved its headquarters from Hampton Bays on the southern coast of Long Island to Riverhead in 1992, in preparation for building a $50 million, full-scale aquarium. Preparations for the complex are still being made, but visitors are welcome to catch a glimpse of things to come at the Aquarium Preview Center, also at the Riverhead location. The Preview Center includes a 22-foot long touch tank with sea stars, urchins, and other local sea-life, as well as interim exhibits focusing on Long Island's various aquatic habitats: bays and marshes, rocky shorelines, freshwater areas, and deep ocean areas.

Involvement Opportunities

Interns can work in any of several areas including education, exhibitry, fishes, research, and animal care. Their intern program is currently under development, and is still fairly flexible. It is typical to find two or three interns at the facility at any given time, usually in the summer. The positions are unpaid, but are usually done for college credit.

Interns are basically woven into one of various programs, with attention given to both the Foundation's needs as well as the intern's goals. In fact, it might not be a bad idea for an applicant to approach them with a proposal for an individual research project which is related to, or compatible with, a program already in place. There is a very good chance that such a proposal would be accepted. A few examples of previous intern activities include local water quality analysis, photo identification of cetaceans, exhibit design, or organizing the volunteer program. Of all the cetacean-related facilities in the U.S., the Riverhead Foundation may very well have the single most flexible internship program, making it an enormous opportunity for the right student (particularly anyone interested in cetacean population dynamics.)

There are over 200 active volunteers at the facility, almost all of which are fairly local residents who donate a few days every month or so. The minimum requirement is one day or ten hours each month, though the more interesting positions will probably command a bit more commitment. There is a "junior" volunteer program for ages 14 to 17, though this group is restricted to non-animal related work. New volunteers are asked to attend a two-hour orientation session after which they are asked to start in such positions as whale watch researcher, exhibit guide, educator, or administrative assistant, among others. After a short time, interested volunteers may ask to switch over to animal care, a position which requires a bit more training. As usual, animal care volunteers spend much of their time in food preparation and cleaning. Becoming a volunteer for a nonprofit group can sometimes cost a few dollars, too - volunteers are requested to become members at a reduced rate of $15, and uniforms must be purchased. The cost is minimal, given the experience they offer in return. Experienced volunteers may be invited to join the Foundation's stranding response team.

Research

Riverhead's scientists conduct numerous research programs, several of which concern cetaceans. The cetacean programs consist primarily of photo identification and biopsy analysis. Both involve getting quite close to pods of cetaceans in the open ocean, so the work can be quite exciting! Photo ID work is a nice alternative to previous methods of population tracking, such as the use of freeze branding, chemical dyes, or invasive tags.

Other Programs

The Foundation has an extensive list of educational programs available for both adults and children. Summer programs offered each year for kids ages 2 to 13 offer an excellent "hands-on" introduction to Long Island's many aquatic habitats. For children over 4, these programs are typically spread over five days and include field trips to local nature sites as well as "home" sessions at the Riverhead facility. Costs range from $50 to $150.

Guided tours of about an hour and a half in duration are available for groups if requested at least a week in advance. Costs vary based on the type of group, but count on $2 for kids and a bit higher for adults. And any junior girl scouts out there looking for an interesting way to earn that elusive Wildlife Badge need look no further - The Riverhead Foundation will work with scout groups to that end.

The Foundation also runs oodles of other programs, including kids' summer programs, whale, seal, and turtle adoption, whale and seal watch trips, adult lecture programs & classes, field trips, cruises, and slide presentations. Memberships are available on an individual or family basis, and include a newsletter and bulletins, free admission to the Center, a nifty decal, and discounts to services. Call or write for the latest information on any of these programs. Internet gurus might also try calling up the Foundation web page, which lists some, though not all, of their educational programs.

New Jersey

New Jersey also possesses a single marine mammal rescue and rehabilitation center which handles strandings for the entire state.

Marine Mammal Stranding Center
P.O. Box 773
Brigantine, NJ 08203
Phone: (609) 266-0538
Fax: (609) 266-6300
http://www.mmsc.org
dolphins@acy.digex.net

Overview

If you're living in New Jersey and looking to donate some of your spare time, maybe you should come a-knocking on the door of the Marine Mammal Stranding Center in Brigantine (just across the way from Atlantic City.) Over a hundred dolphins, seals, whales, and turtles strand, get tangled, or cause false alarms along the New Jersey shoreline each year, mostly in the winter months. Since MMSC is the Northeast Regional Stranding Network's sole representative in New Jersey, they get to investigate every single incident.

Since the group's inception in 1978 they have responded to nearly 1,400 strandings, many of which were successfully rehabbed and sent on their merry way. Every effort is made to bring visiting animals back to a "releasable" level of health, but for those with prohibitive injuries, appropriate lodgings are sought at permanent display facilities. Typical patients include loggerhead sea turtles, harbor seals, and harbor porpoises. A humpback whale stranded one year, and "Chessie" the wandering manatee has occasionally been sighted off the Jersey coast. Ringed, harp, and hooded seals far out of their normal range have shown up as well. Whether more animals have been stranding or people are simply reporting what they see more often, MMSC has been seeing a steadily increasing number of strandings in recent years.

The center is open to the public 7 days a week from Memorial Day to Labor Day. Visitors have the opportunity to learn more about the rescue and rehabilitation process, and tour the Sea Life Educational Center, a 1930's-era coast guard station boat house which offers numerous exhibits and displays. There are twenty-five life-sized replicas of local marine mammals and fish , as well as educational displays to explain the danger of ingesting ocean debris, a common risk for whales, seals, sea turtles and seabirds. A touching exhibit gives visitors a chance to examine various bones and skulls from stranded marine mammals. The center's gift shop provides much-needed funding, so don't forget to pick up a few posters and T-shirts while you're there.

Involvement Opportunities

Although volunteers are in high demand at MMSC, it can be very tricky to find opportunities to assist with the rescue and rehab work. Because the center's staff resources are limited, they have less time to spend on training inexperienced volunteers. This is not to say MMSC doesn't value its volunteers, only that they would sooner use experienced people in the animal areas. Nevertheless, the center provides exposure to the field, and relies a great deal on volunteer assistance in other areas. Volunteers with adequate animal care experience or who have built a history with the center (preferably both) may find themselves recruited for rescue and rehab work. Applicants for volunteer positions are asked to commit to one day each month, and are required to apply in person.

Unpaid college internships are available at MMSC. Interns are required to design a project prior to their acceptance with the assistance of a college professor. The nature of the project is completely up to the student, though it should obviously be compatible with MMSC's mission and facilities. Photo-identification of local dolphin pods is a recommended summer project, since the Center conducts such studies each year. Interns are responsible for their own room and board.

Other Programs

Memberships are available for a mere $15 annually ($25 for families) and gets you a subscription to the center's newsletter. MMSC also offers dolphin-watching cruises in the summer. Call or write for more info.

Delaware

Cetacean strandings are rare in Delaware, and there is no rehabilitation center in the state. Only one response teams exists, run by the state government.

Delaware Division of Fish & Wildlife
PO Box 330
Little Creek, DE 19961
Phone: (302) 739-4782
Fax: (302) 739-6780

The Division of Fish and Wildlife has the only stranding response team within the state of Delaware, which consists of three fisheries biologists. The team responds to anywhere between ten and thirty cetacean strandings each year. Live strandings are rare, but when they occur, the Division will usually enlist the aid of the National Aquarium in Baltimore, and provides no volunteer opportunities for stranding response. In the early summer, the Division occasionally recruits volunteers to assist in a state-wide dolphin survey, which is done in conjunction with organizations like the Virginia Marine Science Museum and the National Aquarium.

Maryland

Home to one of the network's most active rescue and rehab centers, the National Aquarium, Maryland offers a bit more opportunity for involvement than Delaware or Virginia. The Aquarium shares responsibility for stranding response with the state Department of Natural resources and NMFS.

Cooperative Oxford Laboratory
Maryland Department of Natural Resources/NMFS
904 S. Morris Street Extended
Oxford, MD 21654
Phone: (410) 226-5901
Fax: (410) 226-5925
Joyce Evans - Stranding Coordinator

Cooperative Oxford Laboratory is jointly run by the National Marine Fisheries Service and the Maryland DNR. COL personnel are trained to respond to and necropsy beached and stranded carcasses, and assist the National Aquarium in Baltimore in live standings. The COL does maintain a list of local volunteers to assist with dead stranded marine mammals, which occur about a dozen times a year.

National Aquarium in Baltimore
501 E. Pratt St. Pier 3
Baltimore, MD 21202
Phone: (410) 576-3850
Fax: (410) 576-1080

NAIB has one of the most active cetacean rescue and rehabilitation groups in the country, and many volunteer opportunities exist. See page 112 for more information about NAIB.

Virginia

Although both the Virginia Institute of Marine Science and the Virginia Marine Science Museum are involved in stranding response, the Museum is your better bet for volunteering.

Virginia Institute of Marine Science
Gloucester Point, VA 23062-1346
Phone: (804) 642-7000
Fax: (804) 684-7327

The Virginia Institute of Marine Science is a part of the School of Marine Science at the of the College of William and Mary, the second oldest college in the United States. The Institute is a venerable and capable institution, known for its advances in marine research and excellent academic programs. Its involvement in the Marine Mammal Stranding Network is mostly restricted to necropsies and tissue analysis. The Institute often responds to sea turtle strandings, but most marine mammal incidents in Virginia are handled by the Virginia Marine Science Museum.

Virginia Marine Science Museum
717 General Booth Boulevard
Virginia Beach, VA 23451
Phone: (757) 437-4949
Fax: (757) 437-6363
http://www.whro.org/vmsm

VMSM uses a number of local volunteers in its stranding response team. See page 209 for more information about it.

Southeast Marine Mammal Stranding Network

Regional Coordinator
Blair Mase
NMFS Miami Lab
75 Virginia Beach Dr.
Miami, FL 33149
Phone: (305) 361-4284
Fax: (305) 361-1219

Unlike the other stranding networks, the Southeast employs the use of state coordinators to assist the regional coordinator in, well, coordinating things. State coordinators are listed with their home facilities.

North Carolina

This state has two stranding teams, and no rehab facilities. Animals in need of such care would most likely be taken to Virginia, if possible.

National Marine Fisheries Service
Southeast Fisheries Science Center
Beaufort Laboratory
101 Pivers Island Rd.
Beaufort, NC 28516
Phone: (919) 728-8762
Fax: (919) 728-8784

The NMFS Beaufort Lab works in conjunction with the University of North Carolina at Wilmington to respond to nearly one hundred cetacean strandings each year. In addition to NMFS and UNCW employees, the response team relies on volunteer assistance. There are no rehabilitation facilities for cetaceans in the area, and animals in need of such assistance are usually taken out of the state to facilities like the Virginia Marine Science Museum.

University of North Carolina, Wilmington
Biological Sciences
601 S. College Rd.
Wilmington, NC 28403

Phone: (910) 962-7266
Fax: (910) 962-4066
Bill McLellan - State Stranding Coordinator

The stranding response team at the University of North Carolina, Wilmington works directly with strandings in the southern half of the state. It assists the NMFS Beaufort Lab in stranding response to the north. The team responds to all large whale strandings and entanglements from as far north as Delaware and as far south as Georgia.

The University has no long term, live animal care facilities. Animals needing rehabilitation are sent to the National Aquarium in Baltimore or Mote Marine Laboratory on a case by case basis. The Wilmington team does invite local volunteer involvement, though most of its current volunteers are students at the University.

South Carolina

South Carolina is not only without any rehabilitation facilities of its own, but is bordered to the north and south by states with similar situations. Consequently, animals which might otherwise be nursed back to health are sometimes euthanized for lack of a better solution. South Carolina is the only state in the Union to have outlawed cetacean display facilities, which makes the possibility of a future rehabilitation site remote at best. Two government groups, both in the Charleston area, share responsibility for responding to stranded animals.

National Marine Fisheries Service
Charleston Laboratory
217 Ft. Johnson Rd.
Charleston, SC 29412
Phone: (803) 762-8500
Fax: (803) 762-8700

The Charleston Laboratory works with the South Carolina Department of Natural Resources to respond to marine mammal strandings. The Lab has both intern and volunteer programs in place, and is a good place to gain exposure to cetacean research as well as stranding response. While there are no rehabilitation facilities within reach of the South Carolina coast, the Lab has obtained mobile, temporary pools to allow for rehabilitation efforts. The Lab responds to anywhere between ten and seventy strandings per year, about two or three of which are live animals. Volunteer and intern positions are open to anyone, though preference for internships is given to college students.

South Carolina Department of Natural Resources
Charleston Marine Center
PO Box 12559
Charleston, SC 29422
Phone: (803) 762-5000
Fax: (803) 762-5007

The DNR works with the NMFS Charleston office to respond to any marine mammal strandings occurring in South Carolina. They rely heavily on a volunteer network to investigate strandings, and if necessary to call out one of the DNR's three-person teams made up of biologists, technicians, and veterinarians. The DNR handles dozens of strandings each year, and very rarely encounters a live beaked whale or pygmy sperm whale. While volunteers are used by the DNR, they have a more than adequate number, and are not looking to recruit any more.

Georgia

Only the Georgia DNR handles the relatively few cetacean strandings which occur along the state's shores. There is no rehab facility in Georgia, and animals in need of care are usually taken to Marineland in Florida.

Georgia Department of Natural Resources
1 Conservation Way
Brunswick, GA 31520
Coastal Resources
Phone: (912) 264-7218
Fax: (912) 262-3143
Barb Zoodsma - State Stranding Coordinator

There are fewer than forty cetacean strandings along Georgia's coastline each year, and the Georgia DNR handles all of them. It is the only stranding response team in the state, and does utilize local volunteers. Volunteers need only be 18 years of age, and willing to undergo a training program. The DNR encounters 3-6 live strandings per year, which are taken to Marineland in Florida, should they require rehabilitation.

Florida

Sort of the king of the stranding universe, Florida handles more strandings than any other state, and has a proportionately larger number of organizations and volunteers.

Clearwater Marine Aquarium
249 Windward Passage
Clearwater, FL 33767
Phone: (813) 441-1790
Fax: (813) 442-9466
http://www.flaoutdoors.com/wildlife/marine.htm (limited information)

The CMA's primary mission is marine mammal and turtle rescue and rehabilitation, and offers volunteers a chance to get involved. See page 55 for more information about CMA.

Ecological Associates
PO Box 405
Ft. Pierce, FL 34958
Phone: (561) 334-3729
Fax: (561) 334-4925

Ecological Associates is an environmental consulting agency with both government organizations and private industries. A number of employees happen to have been involved with the stranding network prior to the group's inception in 1994, and have formed a response team of their own. The group responds mostly to dead strandings, though they have assisted with live animals in the past. There are no volunteer opportunities with this group; persons looking to volunteer in the Ft. Pierce/Jensen Beach area should contact The Wild Dolphin Project (page 311) or the Florida Department of Environmental Protection in Melbourne (opposite page).

Emerald Coast Wildlife Refuge, Inc.
PO Box 5048
Destin, FL 32540
Phone: (904) 837-4248
Fax: (850) 934-4600
George W. Gray - President

Attempting to fill a gap in the Southeastern Marine Mammal Stranding Network from Destin, FL to as far west as Gulf Shores, Alabama, the Emerald Coast Wildlife Refuge is one of the newest and most enthusiastic members of the network. Currently run out of a number of the staff members' homes, the ECWR has obtained a plot of land in Destin, and hopes to have a rehabilitation facility built by the spring of 1998. The group handles any sick, injured, or abandoned wildlife, and deals with two to three hundred animals per year.

While the ECWR has only dealt with one live cetacean stranding in its three-year existence, it holds a Letter of Authorization from the National Marine Fisheries Service, and its staff has considerable experience in marine mammal stranding response. In addition to building a refuge center, the group hopes to be equipped to rehabilitate cetaceans on-site in the near future.

The ECWR depends heavily on its dedicated volunteer staff, and welcomes help from hard-working people. Springtime is the group's heaviest season, when it receives dozens of orphaned baby birds. For locals along the western half of the Florida panhandle, the ECWR provides the greatest potential for future growth in wildlife rescue and rehabilitation.

Florida Department of Environmental Protection

Similar to the Department of Natural Resources in other states, the Florida DEP is responsible for habitats and wildlife within its state. It is the one member of the Florida stranding network which covers the most area, and has two branches which participate in stranding response. The Florida Marine Patrol often responds first to cetacean strandings, and assesses the situation and informs the appropriate organizations. The Patrol does not generally get involved in actually handling the animals.

That task can fall to the Florida Marine Research Institute, whose biologists and technicians are stationed at both the main compound in St. Petersburg as well as in field offices throughout

Florida. The offices most involved in stranding response are in Melbourne, Jacksonville, Tequesta, and Port Charlotte. There is also a pathology lab down the road from the Institute in St. Petersburg which does a lot of work with samples taken from stranded animals. The various offices are involved in response efforts to different degrees, but most do not make use of volunteers. Contact information for the various offices is given below:

Florida Marine Patrol Phone: (305) 795-2145

Florida Marine Research Institute Phone: (813) 896-8626
Marine Mammal Section
100 8th Ave. SE
St. Petersburg, FL 33701-5095

Jacksonville Office Phone: (904) 448-4300
7825 Bay Meadows Way
Center Building, Suite 200B
Jacksonville, FL 32756-7757

Melbourne Office Phone: (407) 984-4828
1220 Prospect Ave. Suite 285
Melbourne, FL 32901

Tequesta Office Phone: (561) 575-5407
19100 SE Federal Hwy.
Tequesta, FL 33469-1712

Port Charlotte Office Phone: (941) 255-0777
1481-G Market Circle
Port Charlotte, FL 33953

Marine Mammal Pathobiology Lab Phone: (813) 893-2904
3700 54th Ave S.
St. Petersburg, FL 33711

Florida International University
Marine Mammal Rescue Unit
c/o FIU Marine Lab
3000 NE 145th St.
North Miami, FL 33181

Phone: (305) 919-5503
Fax: (305) 919-5684

The Marine Mammal Rescue Unit is a student club, similar to the Marine Mammal Stranding Club at the University of Miami. The Club works in conjunction with local stranding organizations, like the Marine Animal Rescue Society (page 308), and generally assists with about a dozen cetacean and manatee strandings each year. The club is open to faculty and any graduate or undergraduate students at FIU, and often has a membership of more than two hundred students.

The Florida Aquarium
701 Channelside Drive
Tampa, FL 33602
Phone: (813) 273-4020
Fax: (813) 224-9583

The Florida State Aquarium works in conjunction with Mote Marine Lab, Sea World of Florida, the Clearwater Marine Aquarium, and the Florida Department of Environmental Protection to handle marine mammal strandings along Florida's west coast. The Aquarium itself responds to very few strandings, but often provides pathology support for other members of the network. Though opportunities for aquarium volunteers to assist in stranding response are very rare, they do exist, and interested parties should contact the Aquarium's Volunteer Services Department.

Gulf World
15412 Front Beach Road
Panama City Beach, FL 32413
Phone: (904) 234-5271
Fax: (904) 235-8957

Gulf World does not use volunteers to respond to cetacean strandings, but does look for help when actively rehabilitating animals onsite. See page 79 for more information about Gulf World.

Gulfarium
1010 Miracle Strip Parkway
Ft. Walton Beach, FL 32548
Phone: (850) 243-9046
Fax: (850) 664-7858

The Gulfarium responds to cetacean strandings and occasionally rehabilitates animals onsite. See page 80 for more information about the Gulfarium.

Hubbs-Sea World Research Institute
PO Box 691602
Orlando, FL 32869-5397
Phone: (407) 363-2662
Fax: (407) 345-5397

Hubbs/Sea World Research Institute in Florida is an extension of the main office in San Diego, CA. (See page 181.) The Florida branch is headed up by Drs. Nelio Barros and Dan Odell, the latter of which is the Florida state stranding coordinator and also who gathers stranding data from the entire southeastern network. Although Hubbs/Florida is involved in stranding response, it has no opportunities for volunteer assistance.

Marine Mammal Conservancy
PO Box 1625
Key Largo, FL 33037
Phone: (305) 853-0675
Fax: (305) 852-5807
Rick Trout - Director of Husbandry

MMC has accomplished the successful rescue, rehabilitation, and release of 13 stranded marine mammals throughout the Florida Keys since the group's inception in 1995. Its founders were involved in the formation and ultimate dissolution of the Sugarloaf Dolphin Sanctuary. Sugarloaf was, as the MMC puts it, "a false start for reintroduction of captive marine mammals." (For more information on the Sugarloaf debacle, see pages 286-288.) MMC initiated the legal proceedings which wrested two dolphins, Bogie and Becall, from Sugarloaf, and which led to the dolphins' return to native waters.

MMC is currently involved in litigation with APHIS, and hopes to procure and eventually release Luther, Buck, and Jake, and to provide long-term care for Molly; all of whom are ex-Sugarloaf dolphins now under the care of the government. MMC staff hope to reestablish NMFS-permitted, peer-reviewed, and radio-tracked release of appropriate animals into native waters. MMC responds to strandings as a designee under Wildlife Rescue of the Florida Keys (page 311) and does maintain a list of local volunteers to assist in rescue and rehabilitation efforts. "The Florida Bay Dolphin Refuge" is a rehabilitation and educational facility currently under development by MMC, which the group hopes to establish in Key Largo.

Marineland of Florida
9507 Ocean Shore Blvd.
Marineland, FL 32086
Phone: (904) 471-1111
Fax: (904) 461-0156

Marineland responds to cetacean strandings and rehabilitates animals onsite. Volunteers occasionally assist with rehabilitation efforts, but not with rescue attempts. See page 96 for more information about Marineland.

Marine Animal Rescue Society
c/o Florida International University Arts & Sciences
Marine Lab
3000 NE 145th St.
North Miami, FL 33181
Phone: (305) 919-5335
Fax: (305) 919-5684
Craig Pelton - Director

The Marine Animal Rescue Society, one of very few organizations to handle both manatees and cetaceans, grew out of the Marine Mammal Rescue Unit at Florida International University in 1994. The Society's Director, Craig Pelton, and is currently seeking permission from the school to build a rehabilitation facility on school property. The effort, if successful, would provide southeastern Florida with a much-needed venue for sick and injured cetacean care, since the only other viable facility in the area, the Seaquarium, is no longer accepting them. Until the facility is built, the MARS staff will continue to assist the National Marine Fisheries Service in stranding response. The Society currently draws on the University of Miami's Stranding Club and FIU's Marine Mammal Rescue Unit for volunteer assistance, but if the group's plans for growth come to fruition, they may very well accept assistance from other volunteers as well. If you're looking to get involved in stranding response in the Miami area, keep your eye on these guys.

Mote Marine Laboratory
1600 Thompson Parkway
Sarasota, FL 34236
Phone: (941) 388-2451
Fax: (941) 388-4312
http://www.mote.org

Marine mammal rescue and rehabilitation is one of MML's primary missions, and there are many opportunities for volunteers to get involved. See page 190 for more information about MML.

National Aeronautics and Space Administration
Life Sciences Ecological Program
DYN-2
Kennedy Space Center, FL 32899
Phone: (407) 853-3281
Fax: (407) 853-2939

Oops! Wrong career guide! No, wait...this does belong here. Let's see now, I think the dolphins are being sent up to that new International Space Station whatsit. (rustle rustle) Got some notes here somewhere...

Aha! Disregard that last bit. NASA, it seems, has far more gorgeous, prime Florida beachfront real estate than it knows what to do with. (Life is rough all over.) Only a small portion of Cape Canaveral has been developed and put to use, and even after NASA donated a huge portion to the National Park Service as the Canaveral National Seashore, there was still a fair amount of wildlife and seashore within NASA's fences. The Administration set up the Life Sciences Ecological Program to watch over such areas, and one small portion of the Program's duties is stranding response. The group handles few live strandings, and usually enlists the aid of nearby stranding teams when they do. There are limited opportunities for volunteers to get involved with NASA's stranding response efforts. Graduate students interested in manatees or dolphins may sometimes arrange to conduct studies along the undisturbed rivers within Kennedy Space Center.

National Marine Fisheries Service
Miami Laboratory
75 Virginia Beach Dr.
Miami, FL 33149
Phone: (305) 361-4284
Fax: (305) 361-1219
sefscweb@ccgate.ssp.nmfs.gov

The NMFS Miami Lab is the coordinating body for the entire Southeastern Marine Mammal Stranding Network, and also coordinates local response efforts in the Miami area. A slightly more detailed description of the Miami Lab appears on page 267.

National Park Service
Gulf Islands National Seashore
Florida Office
1801 Gulf Breeze Parkway
Gulf Breeze 32561
Phone: (850) 934-2600
Fax: (850) 932-9654

The NPS has a small team of park rangers based in the Pensacola area which are trained, among other things, to respond to marine mammal strandings in the Gulf Islands National Seashore and surrounding waters. The Florida portion of the seashore sees about a dozen or so cetacean strandings each year, most of which are dead. Any live dolphins or whales in need of rehabilitation would most likely be taken to the Gulfarium in Ft. Walton Beach. There are no volunteer opportunities with the NPS. For more information about the National Park Service's involvement in the various stranding networks, see the description of the Gulf Islands National Seashore, Mississippi office on page 314.

University of Miami
Marine Mammal Stranding Club
c/o Marine Science Program
Cox 182
Coral Gables, FL 33124
Phone: (305) 284-2180
Fax: (305) 284-4911

The University of Miami's Stranding Club is very active in the stranding network, and often responds to a dozen or more strandings per year. Originally formed by marine science students, the club is open to anyone with an interest. Members must be able to respond to a call in thirty minutes or less, which I guess makes them the "Dominos" of the stranding world. There are over 80 members in the club, who are divided into four response teams. The teams rotate being on call, giving everyone a chance to participate. During periods of decreased activity, the club often finds other productive tasks to accomplish, and is currently designing a manatee rehabilitation facility.

Wild Dolphin Project
PO Box 8436
Jupiter, FL 33468
Phone: (561) 575-5660
Fax: (561) 575-5681

While the actual dolphin project portion of the Wild Dolphin Project is being conducted outside of the U.S. and Canada (in the Bahamas) and falls outside the scope of this book, the group is headquartered in Florida, and responds to marine mammal strandings in the Palm Beach area. WDP is not as active in training new volunteers as it used to be, but it does still maintain a list of response team members from the local community.

Wildlife Rescue of the Florida Keys
PO Box 5449
Key West, FL 33045
Phone: (305) 294-1441
Fax: (305) 872-0402
Becky Barron - Director of Rehabilitation
bbecbar@aol.com

There are still scattered groups here and there throughout the Florida Keys which help out when they can, but Wildlife Rescue is the only organization in the region which is dedicated to rescuing and rehabilitating cetaceans. They also take in birds, reptiles, and just about anything else which looks sick, abandoned, or injured, except manatees and domestic animals. The group's home compound is on Key West, and is usually home to about a hundred different animals.

Wildlife Rescue has a very interesting method of rehabilitating cetaceans. Rather than transport animals to a holding facility, they bring the facility to Mohammed, and build an open water pen around the animal where it is found. This cuts down on the stress to both cetaceans and humans, and removes a major stumbling block which plagues stranding programs everywhere: the low availability of host facilities.

Over 150 volunteers from throughout the keys are on call for cetacean and turtle strandings, and about twelve of them help out on a weekly basis on Key West. The group cares for over 1,000 animals each year, which can include anywhere from zero to a dozen cetaceans or more. There seems to be no pattern to the frequency of cetacean incidents, though the few pygmy sperm whales they have encountered have all shown up in the summer.

Anyone may apply to help out with the animals, though anybody wishing to get involved in dolphin and whale rescue attempts or rehabilitation efforts must be capable swimmers and in good physical health.

Alabama

There is only one stranding response team in Alabama, which is commensurate with its low number of stranding incidents.

Spring Hill College
Dauphin Island
4000 Dauphin St.
Mobile, AL 36608-1791
Phone: (334) 380-3072
Fax: (334) 460-2198
Dr. Gerald Regan - State Stranding Coordinator

Dauphin Island, a long and skinny mass stretching across the mouth of Mobile Bay, houses the Dauphin Island Sea Lab, a research facility operated by a consortium of 20 Alabama colleges. The consortium has coordinated all marine research in Alabama since 1972. One member of the group, Spring Hill College, hosts the Sea Lab's stranding response team, which handles all cetacean strandings in the state. While most of the team's members are students at Spring Hill, a few local residents have helped out in the past.

Mississippi

Why Mississippi, with almost the least amount of coastline of any state, should have the highest concentration of stranding response teams in the U.S. is beyond me. There are six organizations which participate in the network. One of the facilities, Marine Animal Productions, is equipped to rehabilitate cetaceans.

Marine Animal Productions
Marinelife Oceanarium
P.O. Box 4078
Gulfport, MS 39502
Phone: (601) 864-2511
Fax: (601) 863-3673

Marinelife Aquarium personnel occasionally respond to cetacean strandings, but are more often called upon to rehabilitate animals onsite. Oceanarium volunteers might be able to assist with rehabilitation efforts, though no formal program is in place. See page 92 for more information about Marinelife Oceanarium.

Mississippi Department of Marine Resources
152 Gateway Dr.
Biloxi, MS 39531
Phone: (228) 385-5860
Fax: (228) 385-5917

The MDMR is a regulatory organization responsible for monitoring many of the environmental aspects of the state's coastlines, including the marinelife found within Mississippi's coastal waters. The MDMR serves as the stranding network coordinator for the state. A small stranding response team, coordinated by the DMR and the NMFS, handles about a dozen strandings each year. The group has traditionally encountered one live stranding every other year, on average, though incidents may increase dramatically during red tide incidents. The MDMR does enlist volunteer aid in its stranding response efforts, although federal, state, and local government agency representatives make up the majority of the network.

National Marine Fisheries Service
Pascagoula Laboratory
PO Drawer 1207
Pascagoula, MS 39568-1207
Phone: (601) 762-4591
Fax: (601) 769-9200
Wayne Hoggard - State Coordinator

The Pascagoula Lab coordinates stranding efforts for Mississippi, Alabama, and Louisiana, and itself maintains a response team which responds to roughly one-third of Mississippi's strandings. Live cetacean strandings in the Pascagoula area are typically rare, the most recent of which was a sperm whale in 1993. The Lab does utilize local volunteers in its efforts. Interested people from the eastern coastline of Mississippi can approach the Lab, while western residents might do better to contact the Department of Marine Resources in Biloxi (previous page).

National Park Service
Gulf Islands National Seashore
Mississippi Office
Phone: (601) 875-0821

The Gulf Islands National Seashore extends across portions of Mississippi and Florida, and has rangers in both areas to handle marine mammal strandings throughout the seashore's many islands. The Mississippi section receives even fewer live strandings than in most other areas of the Gulf of Mexico, and has no real opportunities for volunteer assistance.

University of Southern Mississippi
Gulf Coast Research Laboratory
PO Box 7000
Ocean Springs, MS 39566-7000
Phone: (228) 875-2244
Fax: (228) 872-4204

The Gulf Coast Research Laboratory serves not only the University of Southern Mississippi, but provides marine laboratory facilities for about 60 other schools throughout Mississippi and other states. The Lab's Parasitology group and a handful of other technicians, students, and faculty members occasionally assist in local strandings as the opportunity arises. The majority of this assistance is in response to dead strandings, and the Lab does not maintain a list of local volunteers. As a result of its opportunistic involvement, a fair amount of cetacean pathology work is carried out at the Lab, and it is worth noting that a summer course in marine mammology is offered by NMFS scientist Keith Mullin.

Wildlife Rehabilitation and Nature Preservation Society, Inc.
PO Box 209
Long Beach, MS 39560-0209
Phone: (228) 452-9453

The Wildlife Rehabilitation and Nature Preservation Society, Inc. is a non-profit, volunteer organization which rescues, rehabilitates, and releases any indigenous animal that comes through its doors, including squirrels, raccoons, oppossums, all kinds of shorebirds and songbirds, and an occasional alligator. They don't rehabilitate marine mammals, but they are occasionally called in to assist with strandings and oil spills in the Gulf Coast area. The Society is always looking for dependable help from new volunteers, who might assist with animals in the Wildlife Center in Pass Christian, do homecare for baby animals, or help educate the public. Potential volunteers need not be experienced, but must be 18 or older and show up reliably for their shifts.

Louisiana

If God created a hell for stranding response teams, it would be a carbon copy of Louisiana. The state is hot, humid, and encompasses more than 3,000 miles of twisting coastline which consists in great part of impenetrable swampy marshland surrounded by shallow, unnavigable waters. In fact, a full quarter of America's coastal wetlands are found in Louisiana. Since there are few brave or senseless enough to make a habit of sloshing around such areas, most stranding reports come from the helicopters ferrying personnel and supplies to and from Louisiana's offshore oil rigs. When you take into account the time it generally takes for a Louisiana stranding to get noticed and add in a day or so for the pilot to get around to telling someone, you can imagine that once the response team finally wades, hacks, and crawls their way to it they are in for a less than pleasant aromatic experience. (The team members probably won't be starring in any "Sure" commercials either, at that point.)

Since it is so hard to notice and respond to strandings rapidly in Louisiana, response teams rarely encounter live animals, and rehabilitate them even less frequently. Nevertheless, there are two organizations in the state which respond to strandings, one of which takes on volunteers.

Louisiana Department of Wildlife and Fisheries
c/o Louisiana National Heritage Program
PO Box 98000
Baton Rouge, FL 70898-9000
Phone: (504) 765-2821
Fax: (504) 765-2607
Gary Lester - State Stranding Coordinator

The Louisiana Dpt. of Wildlife and Fisheries maintains a team of scientists and field agents to classify, record, and take samples from dead animals. They respond to live animals as well. The Dpt. of Wildlife and Fisheries is a state government organization, and applications can be given to the Baton Rouge office. The Department does not enlist the aid of volunteers, but coordinates stranding responses with the Louisiana Marine Mammal Stranding Network, which does use volunteer assistance.

Louisiana Marine Mammal Stranding Network
c/o Louisiana State University
School of Veterinary Medicine, Clinical Services
Baton Rouge, LA 70820
Phone: (318) 291-5448
Tamara Kley - Assistant State Coordinator
watergirl@linknet.net

Recently incorporated as an independent nonprofit organization, the Louisiana Marine Mammal Stranding Network has centralized efforts at stranding response previously handled by the U.S. Fish and Wildlife Service and the Louisiana Department of Wildlife and Fisheries. The Network still cooperates with the LDWF, but maintains its own stranding response teams, which are staffed in large part by local volunteers. Anyone interested in volunteering is welcome, but they must be willing to participate in the Network's training programs for data collection, necropsies, and live animal handling. The Network developed a relationship with LSU's School of Veterinary Medicine, where it is currently based, and hopes to obtain an administrative and rehabilitation center on campus. The Network rarely deals with live animals, and responds to nearly one hundred strandings each year.

Texas

Texas Marine Mammal Stranding Network
5001Ave. U, Suite 105
Galveston, TX 77551
Phone: (409) 740-4455
Fax: (409) 744-1358
http://www.tmmsn.org
Dr. Graham Worthy - President & State Director
Lance Clark - Operations Coordinator

The TMMSN is made up largely of volunteers from marine laboratories, educators, biologists, toxicologists, and other scientists. The scientific resources at the network's disposal are considerable, and make possible an unprecedented amount of data collection and analysis. All animals released by the program are tracked via satellite, all of which makes the TMMSN far and away the most research-oriented and focused of the stranding networks.

Texas A&M at Galveston hosts the coordinating body for the entire Texas Marine Mammal Stranding Network, and is the institution from which the network grew in the 1970's. As early as 1974, a University professor named Dr. David Schmidly was responding to marine mammal strandings along the Texas Coastline. He represented his state at the first Marine Mammal Stranding Workshop in Athens, GA in 1977. The program is now overseen by Dr. Graham Worthy, also a professor at the University.

Texas A&M, Galveston has donated a holding facility to the program. Though originally designed as an oyster hatchery, this facility has been used to successfully rehabilitate more than seven cetaceans over the past three years. Animals too young to be released are transferred to display facilities equipped to care for them indefinitely, usually Sea World of Texas in San Antonio.

Galveston Region:

Texas A&M University - Galveston
5001Ave. U, Suite 105
Galveston, TX 77551
Phone: (409) 740-4455
Fax: (409) 744-1358
Dan Englehaupt - Regional Coordinator

In the Galveston area, a list of volunteers is maintained by Texas A&M University. Students, faculty, and local residents are all welcome to get involved. The list contains hundreds of names, most of whom are only called upon when live animals require round-the-clock watches. Other volunteers are used to assist in administrative and fundraising tasks.

University of Texas - Medical Branch, Galveston
301 University Blvd
Galveston, TX 77555
Phone: (409) 772-1011
Dr. Dan Cowan - Stranding Network Liaison

While the nearby Texas A&M University handles local volunteers for the stranding network, a small handful of faculty and students from the University of Texas' Medical Branch Pathology Lab respond to strandings as well. Working hard to assist with live animals, the Pathology group is in a unique position to take advantage of dead specimens, taking the opportunity to record data, take samples, and perform necropsies in an attempt to better understand what factors contribute to cetacean strandings. The University's proximity to areas of frequent strandings, its cooperative efforts with other scientists within the network, and the Pathology Lab's own considerable resources combine to form an interesting symbiotic relationship between the Lab and the stranding network: Lab research increases the network's ability to effectively respond to strandings, and provides veterinary, toxicology, and biology students with invaluable experience at the same time.

Corpus Christi Region:

Texas A&M University - Corpus Christi
College of Science and Technology
Dpt. of Physical & Life Sciences
6300 Ocean Dr.
Corpus Christi, TX 78412
Phone: (512) 994-2681
Fax: (512) 994-2742
Dr. David McKee - Regional Coordinator
Linda Price-May - Regional Coordinator

Texas A&M at Corpus Christi brings one of its greatest assets to the table for the stranding network - its students. The school is home to The Texas Marine Mammal Stranding Network Club, which through its very existence bespeaks a particularly strong concern for stranded animals which exists throughout the Texas coastline. The club responds to live and dead strandings under the guidance of its faculty advisors, who are themselves experienced biologists within the stranding network. Bottlenose dolphin calving season in the Gulf is in the spring, which is the period of greatest activity for the Corpus Christi teams.

There are about fifty members in the club - far more than are necessary for the number of strandings in the Corpus Christi area. Not all members choose to assist in actual strandings, but get involved in fundraising efforts and other administrative and support tasks. Anyone is welcome to join the club, including local non-students, though only currently enrolled students may serve as an officer within the group, and must have a 2.25 GPA.

Port O'Connor Region

Texas Parks and Wildlife Department
Port O'Connor Station
PO Box 688
Port O' Connor, TX 77982
Phone: (800) 792-1112
Fax: (512) 983-4404

Rockport Station
702 Navigation Circle
Rockport, TX 78283
Phone: (512) 729-2328
Fax: (512) 729-1437

The Texas Parks and Wildlife Department maintains a number of employees qualified to respond to strandings. The Department operates numerous field stations along the Texas coastline, each staffed by a team of Fish and Wildlife Technicians and Conservation Scientists. These field stations are responsible for keeping an eye on local wildlife and habitats, and making sure nothing unseemly is going on. Team members take water samples, gather invertebrates and fish for testing, and take down harvesting information from fishermen.

There are seven such field stations scattered along the coast, each of which has at least one team member trained to respond to strandings. The field stations at Rockport and Port O'Connor are the most involved in stranding response, and have several individuals constituting response teams. In all cases, the Texas Parks and Wildlife Department encounters far more dead animals than live ones, and generally fill a support role for the University teams when the animals are alive.

There are no volunteers used by the Department's stranding response teams. Stranding response is a very small part of a Department employee's job, though working as a Fish and Wildlife Technician or a Conservation Scientist can provide valuable experience. The Department employs over one hundred of these, and vacancies do occur regularly. Open positions are advertised on bulletin boards at each of the field stations, as well as at Department offices and local Universities.

Port Aransas Region:

University of Texas - Marine Science Institute, Port Aransas
750 Chanelview Dr.
Port Aransas, TX 78373-5015
Phone: (512) 749-6720
Fax: (512) 749 6777
Tony Amos - Research Associate
Andi Wickham - Research Associate

The Marine Science Institute maintains a small list of stranding volunteers from its own staff and students.

A small number of staff and students at the Marine Science Institute are involved in stranding response. The team rarely encounters live strandings, and is more typically called upon to deal with dead animals. Institute volunteers are trained to do basic necropsies, which is to say lab analysis and toxicology procedures are done elsewhere, but it might be up to an Institute volunteer to take measurements, cut samples from a carcass and prepare them for transport.

The Institute's stranding team is also called upon to handle injured and dead turtles and seabirds, and in fact do so relatively often. Live turtles and birds are cared for within the institute's own frugal facilities, and sea turtles are often to be found making a steady recovery there. The Institute does have a somewhat adequate holding facility for small cetaceans, and the Institute is in the process of upgrading and expanding its facilities through the generous contributions of local benefactors, and hopes to be able to increase its capacity for cetacean care considerably in the near future.

South Padre Region

University of Texas - Pan American
Coastal Studies Lab
100 Marine Lab Dr.
South Padre Island, TX 78597
Phone: (210) 761-2644
Don Hockaday - Regional Coordinator

Located very near the Mexican border, the Coastal Studies Lab deals not only with stranded cetaceans but also keeps a lookout for immigrant dolphins trying to sneak across. (Can you tell we writers often work until three in the morning?) In all honesty, the Lab's volunteer stranding team encounters very few live cetacean strandings, having seen only two over the past ten years. Nevertheless, the Lab does maintain a list of stranding response volunteers, some of whom come from the local community.

The Lab is a field station roughly 80 miles from the main campus. The Lab is used most often by the Department of Biology. It provides field trips and laboratory work for a variety of classes. The Lab is too remote to transport live cetaceans to other care facilities. It makes do with its own limited facilities when necessary.

The Lab has no formal volunteer program, and most of its stranding team members are faculty or students. Nevertheless, local residents have assisted in the past, and are accepted on a case-by-case basis.

Northwest Marine Mammal Stranding Network

Regional Coordinator
Dr. Brent Norberg
NMFS NW Regional Office
2725 Montlake Blvd. E.
Seattle, WA 98112
Phone: (206) 526-6733

Washington

Although few cetaceans strand along the Washington coast, the state does get hundreds of sick, injured, and orphaned pinnipeds. Over the years, an occasional gray whale calf, several harbor porpoises, and a few striped dolphins turn up here and there, but very, very few of them survive.

Cascadia Research
218 ½ W. 4th Ave.
Olympia, WA 98501
Phone: (360) 943-7325
Fax: (360) 943-7026

Cascadia's involvement in the network is fairly minimal. They offer few opportunities for volunteers. See page 155 for more information on Cascadia.

Washington Department of Fish and Wildlife
7801 Phillips Rd. SW
Tacoma, WA 98498
Phone: (253) 589-7235
Fax: (253) 589-7180

The DFW's involvement in stranding response is less structured than it used to be. It still responds to such events, but opportunities to get involved are opportunistic at best.

NMFS - National MM Laboratory
7600 Sand Point Way, NE Bldg. 4
Seattle, WA 98115-0070
Phone: (206) 526-4047
Fax: (206) 526-6615
http://nmml01.afsc.noaa.gov/

While NMML personnel respond to strandings on occasion, the Lab offers no opportunities for local volunteers. See page 264 for more on these guys.

The Whale Museum
PO Box 945
62 First Street North
Friday Harbor, WA 98250
Phone: (800) 946-7227
Fax: (360) 378-5790
http://www.whale-museum.com

The Whale Museum responds to many strandings each year, though most involve pinnipeds. The Museum recruits volunteers for the San Juan area. See page 238 for more on this group.

Oregon

Like Washington, Oregon receives far more pinniped strandings than cetaceans. Stranding teams seem to come and go often in the states. At the moment there are four active groups, but this could change at any time.

Oregon Department of Fish and Wildlife
7118 NE Vandenberg Ave,
Corvallis, OR 97330
Phone: (541) 757-4186
Fax: (541) 757-4252

What few volunteer stranding opportunities exist in Oregon can probably best be obtained through the DFW. The department orchestrates much of Oregon's response efforts. Only a dozen or so cetacean strandings have required their attention over the past fifteen years.

Oregon State University
Hatfield Marine Science Center
2030 S. Marine Science Dr.
Newport, OR 97365-5296
Phone: (541) 867-0100
Fax: (541) 867-0128
Dr. Bruce Mate - Research Scientist
bmate@ccmail.orst.edu

Hatfield MSC rarely responds to cetacean strandings, and offers few opportunities for involvement.

Portland State University
724 Harrison
Portland, OR 97201
Phone: (503) 725-4078
Dr. Debbie Duffield

The PSU team has grown less active in stranding response in recent years. Potential volunteers would do better to contact the Oregon DFW on the previous page. (I had to resist the temptation to refer to this place as PU.)

University of Oregon
Oregon Institute of Marine Biology
PO Box 3389
Charleston, OR 97420
Phone: (541) 888-2581
Fax: (541) 888-3250
Dr. Jan Hodder
jhodder@oimb.uoregon.edu

OIMB handles just a few strandings each year, and doesn't enlist volunteer assistance.

California Marine Mammal Stranding Network

Regional Coordinator
Joe Cordero
NMFS SW Regional Office
501 West Ocean Boulevard, Suite 4200
Long Beach, CA 90802-4213
Phone: (562) 980-4017
http://swfsc.ucsd.edu/swr.html

Quite a few cetacean strandings occur along the California coastline. There are half a dozen active groups in the network, many of them rehabilitation centers.

Friends of the Sea Lion
20612 Laguna Canyon Rd.
Laguna Beach, CA 92651
Phone: (714) 494-3050
Fax: (714) 494-2802

The Friends of the Sea Lion is a small rehabilitation facility designed to accommodate seals and sea lions. They cannot rehabilitate cetaceans onsite, though they do occasionally assist with such animals, which are usually taken to Sea World of California. The facility uses a fairly large number of local volunteers, who need only be hard working and dependable to participate. Volunteers participate in nearly all tasks, from animal rescue and care to cleaning, yard maintenance, and sandbagging in rainy seasons. Volunteers are asked to commit to a five hour shift once every week. There are many applicants each year, so expect to be put on a waiting list for a while.

Marine World Africa USA
Marine World Parkway
Vallejo, CA 94589-4006
Phone: (707) 644-4000
Fax: (707) 644-0241

MWAUSA has recently resumed the rehabilitation of stranded marine mammals, though volunteers do not usually come into contact with the animals. The park does not get involved in stranding response, but instead accepts animals from The Marine Mammal Center. See page 98 for more information about MWAUSA.

Northcoast Marine Mammal Center
424 Howe Dr.
Crescent City, CA 95531
Phone: (707) 465-6265
Fax: (707) 465-6265#

The NCMMC is one of the smallest of the California Marine Mammal Stranding Network centers. Situated just 20 miles south of the Oregon border, the facility handles mostly Northern Elephant seals, California sea lions, and harbor seals. Cetaceans strand in this area very rarely, and a live cetacean stranding would most likely be deferred to another group. The Center usually gets 40-60 pinnipeds each year. The elephant seal and harbor seal pup seasons are the Center's busiest months, taking place December through January and March through May, respectively. During less active months, the Center's volunteers focus on training, research projects, and fundraising.

While the Center does receive support from community members, running a marine mammal care center in Crescent City, a fishing community, can sometimes be frustrating. Fishermen often regard pinnipeds as a threat to their livelihood, and more than one sea lion has been brought in with gunshot wounds. The NCMMC has a long track record for successful releases, however, and there is no shortage of volunteers who find the work intensely rewarding.

There are many volunteer opportunities available in animal care and administrative support. Volunteers must be 18 years of age and be willing to commit to a minimum of 4 hours/week during the pup seasons.

Santa Barbara Marine Mammal Center
3930 Harrold
Santa Barbara, CA 93110
Phone: (805) 687-3255

Small but with all the true grit of the most battle-hardened of stranding response teams, the Santa Barbara Marine Mammal Center has handled more than their fair share of rehabilitations. They even managed to free several entangled baleen whales. For example, on April 5th of 1996, the Center responded to scattered reports of a large whale having difficulties swimming along local shoreline. Once volunteer rescuers arrived, they found a young grey whale wound up in nothing less than a monofilament gill net, a synthetic fishing line, an anchor, and several buoys. Once the Center managed to attach flotation devices and a drag chute to keep the whale afloat, I imagine the whole affair must have born an uncanny resemblance to a Gary Larson cartoon.

Wedged between the Channel Islands National Marine Sanctuary to the south and the Los Padres National Forest to the North, Santa Barbara is a marvelous place for viewing protected wildlife. SBMMC handles dozens of pinnipeds and an occasional cetacean each year. The Center relies very heavily on volunteer assistance from the local area, which is a typically gorgeous stretch of southern California coastline. Santa Barbara rests within a long and narrow corridor of developed land from Ventura to San Louis Obispo. The city is framed by the ocean and the Santa Ynez mountains, making the city a beautiful, if intimate, place to live.

About 80 volunteers are on call at the SBMMC. Anyone is welcome to assist, but the Center requests a one full day per week commitment for an entire year. While volunteers are in training for their first six months and are not involved in cetacean rescue operations for their first year or two, they do get a fair amount of animal care experience in addition to general administrative and maintenance tasks.

The Marine Mammal Center
Marin Headlands
Golden Gate National Recreation Area
Ft. Cronkhite, CA 94965
Phone: (415) 289-7325
http://www.tmmc.org/index.html
admin@tmmc.org

Quietly nestled in the Marin Headlands of The Golden Gate National Recreation Area is the rescue, rehabilitation, and research center named, appropriately enough, "The Marine Mammal Center". It rests atop an old and hopefully inert NIKE missile site, which is just one of several remnants of the ex-army post's WWII days.

Overview

TMMC is the largest marine mammal rehabilitation facility in the world. Nestled in the Headlands, the main facility houses the organization's well-equipped hospital, as well as assorted animal holding areas, offices, and a gift shop. There is another gift shop and interpretive center at San Francisco's PIER 39, a popular bayside collection of shops and eateries. Three satellite facilities are maintained in Anchor Bay, Monterey Bay, and San Luis Obispo, stretching the Center's area of operation to nearly one thousand miles of coastline, from San Luis Obispo all the way north through Mendocino county. Each year five to eight hundred marine mammals strand on the beaches within this stretch of coastline. Most of them are pinnipeds, though the Center occasionally takes in sea otters, and they have seen a few cetaceans in past years.* Although stranded cetaceans will most likely be taken elsewhere for housing, TMMC staff is called upon to triage the animal at the stranding site, and to assist in their rescue and rehabilitation.

TMMC, which is a nonprofit organization, has grown considerably since it was formed in 1975. At that time they had perhaps $2000 and half a dozen volunteers to care for the animals which were kept in a bunch of bathtubs somebody managed to rustle up. Any nighttime work had to be done by the headlights of the volunteer's cars. A board of Directors now oversees the operation of the hospital complex as well as the satellite groups. The Center relies almost completely on the financial support of no less than 35,000 members. Over 80% of the center's two million dollar budget goes directly to the animal care and education programs, a fact which is commensurate with the Center's mission statement: "To recognize our interdependence with marine mammals and our responsibility to use our awareness, compassion and intelligence to ensure their survival and the conservation of their habitat."

*In fact, TMMC staff responded to the stranding of Humphrey, a humpback whale who got himself deeply imbedded in the mire of San Francisco Bay in 1990. (I had a similar experience following an incident of unrequited love there in '95, but I certainly don't recall any volunteers scrambling to get ME back on track.) Humphrey was a returning customer for TMMC - they'd already rescued him once before in 1985. Much of the nation watched anxiously as news reporters followed the rescue team's efforts to free Humphrey and prevent him from succumbing to exposure or the crushing weight of his own bulk.

A significant number of the animals they rescue are there because of human activity. The animals suffer from gunshot wounds, pollution, and blockages from marine debris. Quite a few young animals separated from their mothers turn up at the site, as well as stranded adults. Animals which survive and are deemed releasable return to the wild after three to six months. Very young pups may require a longer stay. If an animal cannot be released, but is given a clean bill of health, the staff attempts to place it with a zoo or aquarium.

Intern and Volunteer Opportunities

TMMC has a paid staff of 35, consisting of administrative, educational, and retail personnel, as well as the animal care staff who account for a little less than half the total. There's even an annual veterinary fellowship awarded for vet students. The real opportunities for those just starting out in the field are the volunteer positions. There are over 800 volunteers at the Center, but don't let that dissuade you, as they are often looking for dedicated individuals.

Volunteers work in the gift shop, do mailings, answer phones, staff the booth at the gate, act as educational docents, and of course, assist in animal care. Each area has its own volunteers, though many people find their way into several departments. Some of the more experienced animal care volunteers are even called upon to assist in rescue efforts.

Prospective volunteers for animal care must be at least 18 years old, have a valid driver's license, and must be willing to commit to a certain shift each week. There is one morning and one evening shift each day. Applicants attend an orientation session, arrange a shift with the shift supervisor, and are plugged into the program right away. At first the job will consist of cleaning and food preparation. At the same time they attend courses in basic animal care, and if they stick with it, it isn't long before they learn to do the more hands-on work.

Anyone interested in signing up for an orientation class should call the general number at (415) 289-7325 or the volunteer hotline at (415) 979-4357. If, for some reason, shift availability is low, it might be worth approaching the facility in late winter to early spring. This is when harbor seal pup strandings are on the incline. Harbor seal pups are particularly susceptible to communicable diseases, so the Center creates a separate staff to deal with those animals alone, thereby increasing the demand for volunteers. As with most rehabilitation centers, TMMC staff members do their best to keep the animals in their care from growing accustomed to humans, so contact with the animals may be quite limited. Nevertheless, what you'll learn about animal care and treatment is invaluable, and free.

Research

TMMC's efforts don't stop at animal care and rehabilitation. The Center is uncovering the effects of ocean pollutants on marine mammals, serving as an ecological alert for both human and marine mammals. TMMC medical and science staff work with colleagues from all over the world.

Many of the Center's scientific papers are available through the enthusiastic staff members manning the phones. I was given a list of current research projects at the Center which was nearly as long as my arm, but as the vast majority of their patients are pinnipeds, no ongoing research currently involves cetaceans. Nevertheless, the Center could be an invaluable resource for anyone looking for firsthand experience with marine mammal virology, immunology, or general medical care.

Other Programs

The central facility at the Headlands is open to the public from 10-4 each day except Thanksgiving, New Year's Day, & Christmas. TMMC has recently begun to allocate more resources towards visitors, and it will be interesting to see what sort of exhibitry pops up there in the coming years.

The gift shop at PIER 39 in San Francisco takes advantage of that area's local population of sea lions by posting educational docents to speak about the animals' natural history and characteristics. Now when I say "local population" I'm not just referring a cluster of remote rocks on the horizon, or some nearby secluded inlet. Pier 39 is bordered on the north and south by small harbors of the sort normally used for boat mooring. At some point in the near past, some wayward sea lion must have wandered into one by mistake and found it quite pleasant, because it is now completely overrun by the things. It has to be seen to be believed, or perhaps I should say "heard". If you've ever been around a sea lion, you know what I'm talking about.

Many classes for school children are available both onsite at the Headlands as well as at PIER 39. TMMC educators travel to schools to teach one-day classes. For more information call the Education Department at (415) 298-7330. Group tours may be scheduled by calling the Education Department.

Humpbacks can get to weigh nearly a hundred thousand pounds, provided they stay in the water. When a whale as large as this strands, and loses the water's support, the stress on its body can be tremendous. (This is a very good reason why you so seldom see the larger cetaceans wandering about on land.) For days the rescuers kept Humphrey wet while attempting to dislodge him from the mud. Finally, on the third day, Humphrey shipped out to sea amidst much cheering and jubilation. It was a true success story; one of very few involving one of the larger whales. Still, it seems Humphrey has learned a great deal about what happens when someone like him strands near San Francisco, and if he liked all the attention which was lavished on him during his two visits, we may yet see Humphrey again some day.

Wildlife Emergency Response
PO Box 2
Malibu, CA 90265
Phone: (310) 317-2299
www.wer.org

Wildlife Emergency Response is the newest entry into the field of wild animal rescue and rehabilitation, and is already showing signs of rapid growth, good planning, and capable personnel. It is an all-volunteer outfit, and responds to cetacean strandings as well as handling other marine mammals and terrestrial animals. As of this writing the fledgling organization was still in the process of organizing itself, raising funds, and establishing an animal care facility in Malibu. It is very much in a state of growth, and volunteer opportunities abound, yea, in both animal care and various administrative tasks.

10
Dolphin-Assisted Therapy Programs

Animal-assisted therapy has long been known to be a useful tool for psychologists and social workers dealing with emotionally, mentally, and physically challenged patients. AAT is rapidly gaining widespread professional recognition. It has been known by many other names, and has been practiced since at least the 1970's.

The most common animals used in this kind of therapy are dogs, though nearly any docile, trainable animal can be effective. Rabbits, cats, guinea pigs, regular pigs, and horses are not unusual participants. AAT has been used in patients young and old, and can be used to achieve a wide variety of goals. Regular visits by AAT animals can give terminal patients something to look forward to, and severely disabled patients a new friend to spend time with. Unlike humans, animals accept us for what we are, and pay no attention to our physical appearances.

AAT provides a source of physical interaction to stimulate the use of a patient's arms and hands. Petting dogs and other animals has a calming effect on people, and can even lower a patient's blood pressure. Because the patients usually desire to lengthen time spent with the animals, they increase their endurance in small steps. AAT improves the range of motion in their joints, their perceptive abilities, and even their sense of balance. By enticing patients to try harder almost without their knowledge, AAT can allow people to overcome obstacles they might otherwise find daunting.

This principle can be used with marine mammals as well. AAT with dolphins has been successful in helping young children and other patients overcome irrational fears, self-esteem deficits, and attention-span disorders. Several groups in the U.S. offer the opportunity to incorporate such a program into a patient's treatment, often with impressive results. Some programs, like those offered by the Dolphin Research Center and the Clearwater Marine Aquarium, are designed specifically with younger patients in mind, while others are open to any patients which might benefit from them.

Because AAT, especially as it pertains to cetaceans, is such a relatively new concept, it is still poorly understood. Information about these programs is often obtained second or third-hand. Such information is commonly distorted, giving rise to the misleading notion that dolphins possess some kind of innate healing power. Some literature insinuates that dolphins use echolocation to "beam" into a patient's body in order to initiate some sort of healing process.

Legitimate research on the ability of dolphin echolocation to penetrate an animal's body does not support that theory. The research shows it to be very unlikely a dolphin can obtain any information about the interior of an animal's body in such a manner. It is even more unlikely that an attempt to do so would prompt any unusual healing process.

While dolphin-assisted therapy may for some conjure images of whacked-out ex-hippies, it has been shown that disabled and troubled patients respond well to the process. The procedure is becoming a valuable tool for health professionals looking for a catalyst, a way of reaching out to people who fail to respond to more traditional treatments.

Three DAT programs exist in the U.S., all affiliated with display facilities. This chapter is mostly here as reference material for the inquisitive, but there are a few opportunities within these groups for heath care professionals. Some of the psychologists and social workers involved allow aspiring therapists to participate in internships or volunteer their time.

Dolphin-Human Therapy

13615 South Dixie Highway # 523
Miami, FL 33176-7252
Phone: (305) 378-8670
Phone: (305) 361-3313
Fax: (305) 361-9313
Dr. David Nathanson - President
Diane Sandelin - Internship Coordinator

DHT works on a task-reward principle. About 95% of participants in the program are children between the ages of three and twelve. Once patients achieve a particular task or goal they interact with the dolphins. Motivation is the program's primary goal. Participants work towards improving confidence and attention spans, though specific goals may focus on speech and language skills, motor development, or just about anything else.

DHT is the brainchild of Dr. David Nathanson, who has 20 years of experience as a psychologist working with special needs patients. He was the first psychologist to offer dolphin-assisted therapy in the United States. The program has treated children from 37 states and over 38 foreign countries. It is open to patients with any disability, though the most common include cerebral palsy, Down syndrome, and autism.

In 1997, the program cost $5,800 for daily therapy sessions over a two-week period. Costs for 1998 were not available at press time. The staff encourages family involvement, and welcomes family members and health professionals to observe the program.

DHT offers a limited number of full-time, unpaid internships each year. Internships are open to students and professionals in the areas of medicine, nursing, psychology, special education, and occupational, physical, and speech therapy. Interested parties should contact Diane Sandelin for more information. They are currently pursuing the possibility of opening additional facilities around the world.

The Full Circle Program

c/o Clearwater Marine Aquarium
249 Windward Passage
Clearwater, FL 33767
Phone: (813) 441-1790 x21
Fax: (813) 442-9466
Marianne Klingel - Program Director

The Full Circle Program is a great success story in the area of animal-assisted therapy. The program is so named because the animals themselves are disabled, and in need of long-term care. In addition to Sunset Sam, the Aquarium's single, vision-impaired dolphin, a host of other unreleasable animals gives new hope to children with physical, mental, and emotional disabilities. The turtles, fish, and marine mammals of the Aquarium give children with disabilities something to relate to, since they are living examples of nature's ability to overcome such challenges. At the same time, working with these fascinating creatures helps emotionally troubled kids overcome their fears or attention span disorders. Seeing a child who lost his limbs to meningitis delighting in feeding and caring for an injured turtle is nothing short of inspiring, and I can only hope that such programs can continue to reach out to more children each year.

The Full Circle Program is administered by Clearwater Marine Aquarium therapists, and there are opportunities for CMA volunteers to get involved. In addition, Program Director Marianne Klingel accepts interns each semester. Candidates should be college students pursuing a degree in a related health care field.

Island Dolphin Care

(Dolphins Plus)
PO Box 2728
Key Largo, FL 33037
Phone: (305) 451-5884
Fax: (305) 451-3710
Dr. Deena Hoagland - MSW, LCSW
Dr. William Shannon - PhD, School Psychologist

Island Dolphin Care is a not for profit organization formed to help troubled children. The program is run from the Dolphins Plus facility in Key Largo, Florida, roughly an hour and a half from the Miami International Airport. (See page 72 for more on Dolphins Plus.)

Typical therapy packages last one week. Families arrive for orientation on Sunday. During the following week participants engage in daily interactions with the animals according to a structured regimen custom-designed by the therapy team. In addition to helping overcome irrational fears, disability-related depression, and attention span disorders, Island Dolphin Care believes the experience can help reinforce or even re-establish troubled relationships. Therapist Deena Hoagland has a unique perspective on such situations, having gone through the same process herself with her son, Joe.

Additional dolphin swims for other family members can be arranged. So can occupational, speech, and physical therapy sessions for program participants. Psychological, intellectual, and developmental assessments are also available, as are therapeutic massages and respite care services. Costs for a one-week program run at just under two thousand dollars, which includes daily dolphin sessions and four classroom sessions.

11
Interest Groups

My use of the term "interest groups" in this book includes animal rights activists and spiritualists which focus specifically on cetaceans. There are two groups of both kinds. I mean to imply no link between these two types of organizations, I just couldn't justify assigning each its own chapter.

The term "animal rights groups" may be misleading. Many organizations in Chapter Six seek to advocate the rights of wild animals. The main difference lies in conservation groups' efforts to protect entire species, whereas the Dolphin Alliance and The International Dolphin Project choose to focus their efforts on animals in display facilities.

As for the spiritual groups, I've decided to let them speak for themselves.

Dancing Dolphin Institute

PO Box 959
Kihei, HI 96753
Phone: (808) 573-4235
Fax: (808) 874-5888
http://www.maui.net/~dolfyna
dolfyna@maui.net

The Dancing Dolphin Institute is one of the up and coming players in the "cetacean spirituality" arena. Reprinted here is the group's mission statement:

> "The Dancing Dolphin Institute (DDI) was founded in 1994 with the twin aims of fostering research in interspecies communication and consciousness and exploring its relevance for human affairs. We see individuals awakening to a more profound understanding of themselves in connection with cetaceans and the ocean.
>
> "The Institute will seek to find creative ways to reinforce the rainbow bridge that exists between cetaceans and humans. To broaden awareness of our mission, we will report on people, projects, trends and ideas that embody this vision. We will present seminars with dolphin and whale authorities, healers, and new-consciousness speakers. As the Institute grows in strength and stature, this will be a signal to the world. We are a world awakening!
>
> "Fostering awareness, appreciation, and understanding of the cetacean species (dolphins and whales), the Institute will encourage better communication between our two species. This will be accomplished in part through our educational programs for children, families, and the general public. The Institute will gather information and feedback through these programs that will assist us in our research into interspecies communication and therapeutic applications. Healing the Planet at a broad-based level will be accomplished through the Institute introducing individuals to the gifts of wisdom and consciousness that dolphins and whales offer us.
>
> "The vision is to create a Cetacean Healing Center and Rejuvenation Spa in Maui, Hawaii. The retreat complex is designed to create and ambiance of unity, collaboration, cre-

ativity, inspiration, harmony, and relaxation in order to initiate the transformation of the human species."

The Institute's sister organization, Dancing Dolphin Press, is a company designed to support the Institute through sales of books, audio tapes, videos, and other products. Videotapes include Maui Meditation, Africa: Earth Energy, Paradise Blue, and Canyon Sanctuary, all for $19.95. "1 Heart, 1 Home" T-shirts are offered for $20 each, and a hand-painted dolphin or whale silk scarf will run you $75. You can order any of these products by sending a check to Dancing Dolphin Press at the Institute's address. Don't forget to toss in three bucks for shipping and handling.

Shipping and handling. You know, everyone always says that. Just what exactly constitutes shipping, and what is its distinction from handling? What if I don't really want some stranger handling my merchandise? I mean, I don't even know what this hypothetical shipping agent, who might be a big, greasy, health-hazard of a man for all I know, has in mind for package "handling". Can I get that taken off the bill? Look, just ship it, will ya?

Dolphin Alliance

PO Box 510273
Melbourne Beach, FL 32951
Phone: (407) 951-1301
Fax: (407) 951-1301
http://www.envirolink.org/arrs/ahimsa/tda/index3.html

The Dolphin Alliance is a grassroots organization of people concerned with the well being of dolphins and whales, both in the wild and on display. The Alliance has five main goals: "to end wild captures of cetaceans in U.S. waters, to rehabilitate and release all cetaceans currently in display facilities, to educate the public about the problems dolphins and whales face, to support the protection of marine habitats, with an emphasis on the Indian River Lagoon in Melbourne, Florida, and non-invasive research on dolphin communication and behavior in the field."

The Dolphin Alliance credits itself with the passing of numerous protective laws and regulations within the state of Florida, and has organized many letter-writing campaigns, protests, and petitions. The Alliance relies heavily on volunteers to power such efforts, and welcomes assistance from interested parties. Memberships are also available for about $20 per year, which includes a subscription to the Alliance's newsletter, Wild About Dolphins.

Human-Dolphin Interaction

P.O. Box 4277
Waianae, HI 96792-1932
Phone: (808) 696-4414
Fax: (808) 696-4454
Dolphins4U@aol.com
Terry Pinney Mackenzie - Director

The following is an excerpt from the Aquathought website:

(http://www.aquathought.com/idatra/symposium/96/pinney.html)

> "Terry Pinney MacKenzie has come forward to volunteer her unique telepathic abilities to act as a bridge (or sensory link) between two worlds, i.e., the Scientific and Therapeutic Communities and the dolphins. Terry has developed an ability to telepathically communicate directly with dolphins, tapping into her natural clairaudient, clairsensient, and clairvoyant sensibilities to facilitate inter-species communications. The results are extraordinary.
>
> "The dolphins have had this information for millenniums. But now, we as humans, are capable of opening up to experience what they are communicating to us. We all have this capacity to understand, and Terry is simply one of many who have come forward to share this information with the scientific community."

International Dolphin Project

P.O. Box 2599
Ventura, California 93002
Phone: (805) 655-5523
Fax: (805) 648-1610
http://www.envirolink.org/arrs/ahimsa/idp/index.html
idp@rain.org
Ric O'Barry - Director Of Rehabilitation and Release

The IDP is a group dedicated to releasing captive dolphins back into the wild. This group seems to have become less active in the U.S. in recent months, though it has not disappeared. See page 286 for more information on IDP director Ric O'Barry.

Part Four

Appendices

- ►Schools With Marine Mammal Programs
- ►Books You Should Read
- ►Organizations Listed By State

Appendix A
Schools With Marine Mammal Programs

Here's a list of colleges and universities which offer entire courses of study in marine mammology. Most of these are Master of Science programs, though they may offer one or more marine mammal-related classes to undergraduates as well. Many programs are cross-referenced with other descriptions in the book.

California State University
Moss Landing Marine Laboratories
Ornithology and Mammology Lab
PO Box 450
Moss Landing CA 95039-0450
http://color.mlml.calstate.edu:80/www/mlml.htm
Phone: (408) 633-7261
Fax: (408) 633-0805
(see page 188)

College of the Atlantic
105 Eden St.
Bar Harbor, ME 04609
inquiry@ecology.coa.edu
Phone: (800) 528-0025
Fax: (207) 288-4126
(see Allied Whale, page 151)

Moorpark College
Exotic Animal Training & Management Program
7075 Campus Rd.
Moorpark, CA 93021
http://www.vcccd.cc.ca.us/eatm/eatminfo.html
eatm@sunny.vcccd.cc.ca.us
Phone: (805) 378-1441

Moorpark college has the distinction of being the only college in the U.S. to offer courses on the care and training of exotic animals, including marine mammals. EATM is a two-year college program, offering majors in General Exotic Animal Training and Management, Animal Behavior Management, and Wildlife Education. Moorpark admits only fifty students into the program each year, and not all complete it. The college places 80% of its EATM graduates into zoo, aquarium, and marine park jobs. Moorpark can provide an excellent pathway for aspiring dolphin trainers. As discussed on page 40, it may limit potential career growth unless supplemented with a four-year degree. Application materials which outline the program in greater detail cost $3.00 and a college catalog costs $4.00. Order either by sending a check payable to EATM to the address above.

San Diego State University Phone: (619) 594-5200
Psychology Department
Cetacean Behavior Lab Phone: (619) 594-5649
San Diego, CA 92182 Fax: (619) 594-1332
R.H. Defran - Director
http:www.sci.sdsu.edu/cbl/cblhome.html
rdefran@sunstroke.sdsu.edu
(see page 163)

Texas A&M University at Galveston Phone: (409) 740-4718
MM Research Program Fax: (409) 740-4717
4700 Avenue U, Bldg 303
Galveston, TX 77553-1675
Email - mmrpinfo@mmrp.tamu.edu
Dr. Bernd Wursig - Director
http://www.tamug.tamu.edu/~mmrp/
(see page 184)

University of California, Santa Cruz Phone: (408) 459-2464
Natural Sciences Division Fax: (408) 459-4882
Institute of Marine Sciences
A315 Earth and Marine Sciences Building
Santa Cruz, CA 95064
also
Long Marine Lab/UCSC Phone: (408) 459-2883
100 Shaffer Rd. Fax: (408) 459-3383
Santa Cruz, CA 95060
(see page 89)

University of Hawaii Phone: (808) 236-7401
The Hawaii Institute of Marine Biology Fax: (808) 236-7443
P.O. Box 1346
Kaneohe, Hawaii 96744
http://www.soest.hawaii.edu/HIMB/himb.html
(see page 81)

The Following is a list of colleges which offered at least one marine mammal course in 1997:

College of DuPage
Duke University School of the Environment
Eckerd College
Grossmont College
Nova Southeastern University
Orange Coast College
Oregon State University
San Francisco State University
San Jose State University
Spring Hill College
University of Alberta
University of British Columbia
University of Florida
University of Georgetown
University of Guleph
University of North Carolina, Wilmington
University of Oregon
University of Rhode Island
University of South Florida
University of Southern Mississippi
University of Victoria
University of Washington, Seattle
Western Illinois University
Memorial University of Newfoundland

Appendix B
Books You Should Read

Here are the three best books to check out for general information on cetaceans.

Whales, Dolphins, and Porpoises
Edited by Sir Richard Harrison and Dr. M.M. Bryden
Published by Facts on File Publications, New York
This is hands-down the best introductory coffee table book on cetaceans anywhere. The photography is gorgeous, and the text covers everything from evolution to whaling. Don't confuse this with the National Geographic book by the same name, which looks very similar.

Whales, Dolphins, and Porpoises
By Mark Carwardine
Published by Dorling Kindersley, New York
Many books bear this name. This one is a pleasure to look at, thanks to the beautiful illustrations by Martin Camm, which appear on every page. It does a fantastic job of categorizing, describing, and illustrating every family and species of cetacean known today. It's an invaluable reference text, and just plain fun to leaf through.

The Lives of Whales and Dolphins
By Richard C. Connor and Dawn Micklethwaite Peterson
Published by Henry Holt and Company, New York
Brought to you by the American Museum of Natural History, this book goes into a little greater detail about cetacean biology and behavior than the other two books.

Here are the three best books to learn a little more scientific information about cetaceans. They reprint the results of studies which normally appear in scientific journals.

Cetacean Behavior
Edited by Dr. Louis Herman
Published by Roger E Krieger Publishing Co., inc., Malabar, FL
This covers a wide range of behavioral topics, and will give you a good idea of what scientific writing is all about.

Dolphin Societies: Discoveries and Puzzles
Edited by Karen Pryor and Kenneth Norris
Published by University of California Press, Berkeley

Another excellent collection of scientific papers brought to you by two of the biggest names in cetacean research.

Self-Awareness in Animals and Humans
Edited by Sue Taylor Parker, Robert Mitchell, and Maria Boccia
Published by Cambridge University Press, Cambridge

Only two papers in this collection of studies on self-awareness pertain directly to cetaceans, but they clearly show dolphins know what a mirror is.

Here are the three best books for looking at cetaceans through the eyes of experienced researchers and naturalists.

Among Whales
By Roger Payne
Published by Charles Scribner's Sons, New York

Dr. Payne's efforts in conservation and research are nearly unparalleled. His writing is insightful and a pleasure to read.

Dolphin Days: The Life and Times of the Spinner Dolphin
By Kenneth Norris
Published by W.W. Norton & Co., New York

Another volume from Dr. Norris, this time a personal accounting of his days at sea with some of its more interesting inhabitants. Also well-written and easy to read. It addresses not only behavioral and biological aspects of spinner dolphins, but their ultimate fate at the hands of man as well.

The Presence of Whales
Edited by Frank Stewart
Published by Alaska Northwest Books, Anchorage

An outstanding collection of essays from over two dozen of our most accomplished cetacean scientists, naturalists, and conservationists. See who else shares your interest in these animals, and why.

This book is a must for anyone interested in animal training. It won't substitute for a psychology degree, but it's a terrific start. There are other books by Karen Pryor as well.

Don't Shoot the Dog! The New Art of Teaching and Training
By Karen Pryor
Published by Bantam Books, New York
It's the best introduction to the techniques used by dolphin trainers everywhere. What more can I say?

If you're interested in whale watching on the West Coast, or anywhere for that matter, this is a great resource:

West Coast Whale Watching
By Richard C. Kreitman and Mary Jane Schramm
Published by Harper Collins West, New York
West Coast residents and visitors will most benefit from this well prepared book, but it contains a very good introduction to cetacean natural history and whale watching in general.

Pick these up to learn about current issues in cetacean research, conservation, and politics.

Annual Report to Congress
By the Marine Mammal Commission
If you're looking for a job in cetacean display, research, or conservation, you really should read this.

Dolphins, Porpoises, and Whales
By the IUCN
This gives you a good idea of what issues cetaceans are facing world-wide. Also a good idea for career-seekers.

Appendix C
Organizations Listed By State

Display Facilities:

California

Connecticut

Florida

Hawaii

Illinois

Research Organizations:

Washington

British Columbia

Quebec

Conservation Groups:

CHECK THE OMEGA WEBSITE EVERY SO OFTEN FOR UPDATES AND ANNOUNCEMENTS OF FUTURE REVISIONS AND OTHER PRODUCTS!

HTTP://WWW.OMEGALAND.COM

If you have any questions or comments about this book, mail them to:

Omega/Publishing Division
429 W. Ohio #123
Chicago, IL 60610

or email to:

questions@omegaland.com

or fax them to:

(312) 663-5254

Afterword

Well, there it is. I sincerely hope you find this book practical, helpful, and at least mildly fun to read. I only wish I could have crammed even more information in, but after you go through seven or eight absolutely final, no-way-you-can-miss-this-one deadlines, the time comes to wrap things up as best you can.

Never let anyone dissuade you from pursuing your dreams. A passionate desire which harms no-one and is even remotely plausible should not be sacrificed in the name of convention or established protocol. Learn as much as you can, experience as much as you can, and surround yourself with whatever and whoever makes you happy. Life's just too damn short to do anything else.

Thomas Glen
University of Chicago USITE Facility
Halloween night, 1997

Theebida theebida theebida that’s all, folks.